CAMBRIDGE LIBRARY COLLECTION

Books of enduring scholarly value

Earth Sciences

In the nineteenth century, geology emerged as a distinct academic discipline. It pointed the way towards the theory of evolution, as scientists including Gideon Mantell, Adam Sedgwick, Charles Lyell and Roderick Murchison began to use the evidence of minerals, rock formations and fossils to demonstrate that the earth was older by millions of years than the conventional, Bible-based wisdom had supposed. They argued convincingly that the climate, flora and fauna of the distant past could be deduced from geological evidence. Volcanic activity, the formation of mountains, and the action of glaciers and rivers, tides and ocean currents also became better understood. This series includes landmark publications by pioneers of the modern earth sciences, who advanced the scientific understanding of our planet and the processes by which it is constantly re-shaped.

Life of Sir Roderick I. Murchison

Sir Roderick Impey Murchison (1792–1871) was an influential Scottish geologist best known for his classification of Palaeozoic rocks into the Silurian system. After early military experience in the Peninsular War, he resigned his commission; a chance meeting with Sir Humphrey Davy led him subsequently to pursue a scientific career. *The Silurian System*, published in 1839, was a highly influential study, which established the oldest contemporary classification of fossil-bearing strata. Murchison was appointed President of the Royal Geographical Society in 1843. These volumes, first published in 1875, use information taken from Murchison's private journals and correspondence. Archibald Geikie (1835–1924) provides a detailed account of his mentor's life and work in the context of geology as a developing science in the early nineteenth century, and provides a fascinating insight into the life and work of this eminent Victorian geologist. Volume 2 describes his later life, from 1843 to 1871.

Cambridge University Press has long been a pioneer in the reissuing of out-of-print titles from its own backlist, producing digital reprints of books that are still sought after by scholars and students but could not be reprinted economically using traditional technology. The Cambridge Library Collection extends this activity to a wider range of books which are still of importance to researchers and professionals, either for the source material they contain, or as landmarks in the history of their academic discipline.

Drawing from the world-renowned collections in the Cambridge University Library, and guided by the advice of experts in each subject area, Cambridge University Press is using state-of-the-art scanning machines in its own Printing House to capture the content of each book selected for inclusion. The files are processed to give a consistently clear, crisp image, and the books finished to the high quality standard for which the Press is recognised around the world. The latest print-on-demand technology ensures that the books will remain available indefinitely, and that orders for single or multiple copies can quickly be supplied.

The Cambridge Library Collection will bring back to life books of enduring scholarly value (including out-of-copyright works originally issued by other publishers) across a wide range of disciplines in the humanities and social sciences and in science and technology.

Life of
Sir Roderick I.
Murchison

Based on his Journals and Letters

VOLUME 2

ARCHIBALD GEIKIE

CAMBRIDGE
UNIVERSITY PRESS

CAMBRIDGE UNIVERSITY PRESS

Cambridge, New York, Melbourne, Madrid, Cape Town,
Singapore, São Paolo, Delhi, Tokyo, Mexico City

Published in the United States of America by Cambridge University Press, New York

www.cambridge.org
Information on this title: www.cambridge.org/9781108072359

This edition first published 1875
This digitally printed version 2011

ISBN 978-1-108-07235-9 Paperback

LIFE

OF

SIR RODERICK I. MURCHISON.

LIFE

OF

SIR RODERICK I. MURCHISON

BART.; K.C.B., F.R.S.; SOMETIME DIRECTOR-GENERAL OF THE GEOLOGICAL
SURVEY OF THE UNITED KINGDOM.

BASED ON HIS JOURNALS AND LETTERS

WITH NOTICES OF HIS SCIENTIFIC CONTEMPORARIES
AND A SKETCH OF THE RISE AND GROWTH OF
PALÆOZOIC GEOLOGY IN BRITAIN

BY ARCHIBALD GEIKIE, LL.D., F.R.S.

DIRECTOR OF H.M. GEOLOGICAL SURVEY OF SCOTLAND, AND MURCHISON PROFESSOR OF GEOLOGY
AND MINERALOGY IN THE UNIVERSITY OF EDINBURGH.

IN TWO VOLUMES—VOL. II.

Illustrated with Portraits and Woodcuts

LONDON
JOHN MURRAY, ALBEMARLE STREET
1875.

CONTENTS OF VOL. II.

CONTENTS OF VOL. II.

LIST OF ILLUSTRATIONS IN VOL. II.

CHAPTER XVII.

JOURNEYS TO COMPLETE THE WORK ON RUSSIA.

RELEASED from the trammels of office, Murchison began to prepare for an early start to the Continent. His Russian work needed much additional confirmation and elucidation from adjacent countries, and he resolved to perfect it, as far as possible, by further research in the field. In the midst of these preparations a small but useful piece of work was begun, which is referred to in the subjoined letter :—

"*24th February* 1843.

"MY DEAR SEDGWICK,—The enclosed is part of a very *wee* map of England about to be published by the Society of Useful Knowledge, and which I have (after promising to do something for three years) at last thrown into form. You perceive, for its size, that nothing very precise can be attempted, and all I wish you to do is to mark with · your pencil, in Wales, the tracts which are *igneous,* and those which are *pure slates without fossils*, putting a (X) on the fiery dogs. No name will be appended to it, and no reputation is involved.[1]

"In a day or two you shall have the slips of the

---

[1] The late Dr. S. P. Woodward had the chief share, it is understood, in the preparation of this map for the engraver.

Palæozoic parts of my discourse, which I wish you to look over, prune, and correct.[1] . . .

"We are now all tranquil again, or rather fighting away at our old concerns, and I am deep in printing *de omnibus.*

"Hoping your London let off did you no harm, believe me," etc.

A few days later, having meanwhile received Sedgwick's reply and assistance, he again writes :—

"Thanks for your pencillings and remarks, from which I shall be able to cobble up something better in a small way than anything which has yet appeared in reference to the older rocks.

"I have had a most agreeable letter from old D'Omalius d'Halloy, who, acting as he always does, like a lover of truth, informs me that he is going to publish a new edition of his work, in which he is going to swamp his own anthraxiferous and slaty children,[2] and adopt our classification of Carboniferous, Devonian, and Silurian for the Belgian countries. There's a triumph for us, my friend!"

A good many years had now passed away since Mrs. Murchison took her part in a Continental excursion. It was arranged that she should accompany her husband this year, settling down at some central place, and leaving him to make more distant and laborious forays by himself. They started in April, and went first to Paris. Murchison has left some reminiscences of this time. "At the Institute and elsewhere I had arguments with Elie de Beaumont, when I found that he disbelieved the statements of Sedgwick and myself in our tour of 1839, viz., that in Bavaria there existed

---

[1] See *ante*, vol. i., note on p. 380.
[2] See his Table, quoted *antea*, vol. i. p. 178, *note.*

an entire conformity between the Carboniferous Limestone
and the underlying Devonian. This phenomenon did not
suit the theory of the man of authority who was to become
a Senator of Napoleon the Third, and Secretary of the Institute. I never had any serious dispute with De Beaumont.
But as he settled his belief on certain data known to him
then, and formed his theory, which would not tally with
the new discoveries, which he ignored, I necessarily went
forward, and so offended him. My excellent friend De
Verneuil and others have shared the same fate. In those
days, however, we were on very friendly terms, and, as on
former occasions, he gave me a dinner at the Palais Royal.
D'Archiac was then rising fast to the eminence as a geological writer and sound reasoner (combining palæontology
and field geology) which he has now attained (1865). On
the other hand, D'Orbigny, who was a dashing palæontologist, and too fond of rapid identifications, though he made
beautiful collections of fossils, was evidently on the wane.

" It was on the occasion of this visit to Paris that I was
presented to Louis Philippe, and had a long conversation
with him. M. Guizot, who was then Prime Minister, and
who, when he came to London as ambassador, had dined
with me at the Geographical Club (he being then President
of the French Geographical Society), asked me to one of his
'grands dîners,' at which the Chancellor, Pasquier, and
several great folks were present. On the following day I
received a letter, evidently the work of Guizot, from the
Aide-de-camp au Roi, saying that his Majesty would be
happy to receive me at the Tuileries at 12 o'clock on the
following day, and mentioning the gate at which I was to
enter. Driving thither, in full uniform, and my *Silurian*

System in hand to present, my approach to the King's
saloon showed me how the Roi citoyen might at any moment
be disposed of in case of an insurrection.   My carriage drew
up close to a side-door, which at once opened into a small
room in which several clerks were writing, as in a counting-
house, one of whom told me to sit down.   Presently a livery
servant (none of the cleanest) appeared, and asked me, ' Est-
ce que Monsieur vient voir le Roi ?' and then told me that
the King would soon see me.   After waiting a few minutes,
he returned, saying, ' Le Roi vous verra !' and opening a door
from this writing-shop, there was the King, who literally
seemed to open the doors for me himself.   No chamberlain,
no officer, not even a sentry between the King and an arch-
way by which all the people passed.

"Louis Philippe was most affable and kind in his
manner.   He made me sit down with him in a bay-window
facing the Carrousel, and begged me to unfold my maps and
explain them, saying that he was profoundly ignorant of my
science.   He nevertheless talked of the great coal formations he had seen in the United States, and, in alluding to
his travels in Norway, related the following story :—' I was
one day (said he) standing on the sea-shore, and gazing at a
ship in the offing, when an old pastor of the country, eighty
years of age, who was near me, exclaimed, "You only look
at the sea, sir, but you do not see what is under your feet !"
On doing so, I found that I was standing on gravel and sea-
shells, a little above high-water mark.   The old clergyman
then continued :—"When I was young the sea washed these
shells, but now it never reaches them ; and so you see we
believe that our land is rising !"   From that moment I conceived (added the King) that the earth is always swelling

out as a balloon when it is being inflated; but pardon me if
my theory is foolish and untenable !'

"In the course of the interview the same lacquey returned, saying gruffly, 'Monsieur Guizot, votre Majesté!'
on which the King exclaimed, 'Ah! Monsieur Guizot is my
master, and I must go to him; but I would wish you to do
me the favour to wait here a few minutes. I will settle
matters with him, for I want to return and go on with this
interesting talk."

"The bright-eyed Citizen King kept his promise, and
was both entertaining and pleasing in a long subsequent
talk. A day or two afterwards he sent me a large gold
medal, with his head on one side, and on the reverse, 'A
M. Murchison, de la part du Roi.' In short, I had every
reason to be pleased with the courteous and gracious reception I received at the hand of Louis Philippe; but I came
away with the impression that he wanted that dignity and
reserve which imposes upon the French people, and had put
himself into a false position by the absence of all state
entourage, without which no one, whether king or emperor,
can rule France."

From Paris the travellers journeyed into Rhineland, and
thence parting from Mrs. Murchison, the geologist struck
eastward to increase the collection of materials for the geological map of Russia and the surrounding countries which
it was proposed should accompany the letterpress of the
large work on the geology of that Empire. "Leaving my
wife," he says, "at Baden, who was to meet me again in the
Alps, I went on by Carlsruhe to Heidelberg, conferring with
Walchner at the former, and was well occupied in good old
Leonhard's museum at the latter place. . . .

"Dear old Leonhard, with his pipe and his little bandy-legged dog 'Tegel,' was a fine specimen of a polished German Professor. Igneous to the backbone, for he even believed that rock-salt, gypsum, and hæmatitic iron were produced by intense heat and fusion, he admitted, but could not explain, the difficulty I had had in examining the Ural Mountains, viz., why the iron-ore which is in contact with eruptive rocks is the most magnetic? On another count his disbelief in the metamorphism of great mountain masses of gneiss and mica-schist was founded solely on his own minute researches, in which he had never seen igneous effects extend more than a few feet or yards into sedimentary strata beyond the point of contact."

Before trusting himself to the unknown wilds of Poland and the Carpathians, Murchison turned aside to pay a short visit to Berlin, with the view of renewing his acquaintance with the geologists there, and gaining information regarding Russia as well as the scene of his purposed new labours. His journal contains some reminiscences of Humboldt and Court life which may be quoted :—"From Berlin I went to Potsdam to see Humboldt, on Sunday the 28th of May, after an absence of two years. On this occasion I had brought with me the sketch of the Geological Map of Russia in Europe and the Ural Mountains, and consulted him on all points connected therewith, to profit by his advice and his additions. He went into some detail on various points. 'But first,' he said, 'the moment I mentioned to the King that you were coming here, he begged you would dine with him on this, his only day for receiving foreigners. Unluckily I had no time to let you know the King's wishes, and of course you have come without your dress clothes ; so the King's views cannot be met. But let me,' said the kind old philosopher, 'go immediately to his Majesty and we will see what can be done.'

"In the Baron's absence I was arranging my maps, and he presently came back, saying, The King, regretting very much you cannot dine with him, wishes to see you at once. His Majesty is sitting for his portrait in the dining-room (on account of the light), and he will thus have an opportunity of talking with you—only put off your great-coat.' This being done, and taking my hat in my hand, and putting my work on the Rhenish Provinces into the hand of Humboldt, I walked with him along a corridor, through the great coach-roofed *salon d'entrée*, through a little anteroom, with dinner set for the officers, and thence into the dinner-room, in which I was to have dined if I had had a pair of black pantaloons and a coat with me!

"I had, however, so friendly a reception that it was worth many dinners. Passing behind the screen to go to the window in which the King was sitting to the artist, his voice was at once heard. ' Ist das der Murchison ?' and in a second I was before him. On his countenance was the same complacent smile, in his small blue eye the same kind, cheering, and intelligent twinkle, which left its impression upon one in the saloons of the Duchess of Sutherland, at the good deceased Duke of Sussex's *déjeûner* at Kensington, and last, not least, when the King honoured us by becoming a geologist in Somerset House.[1]

"After all sorts of preliminary inquiries he went on at once

---

[1] The King of Prussia, on the occasion of his recent visit to London, had been admitted an Ordinary Fellow of the Royal and Geological Societies.

to home and London questions—the building of the Houses
of Parliament, Nelson's Monument in Trafalgar Square, the
statue for its top, who it was that admitted him into the
Royal Society, etc. The conversation was then turned by
Humboldt to Russia, and the King expressed his surprise
that I had been so great a traveller, on which I had only to
say that I was a pigmy alongside of Baron Humboldt. The
Rhine-works were then alluded to and Prussian geology." . . .

"On afterwards talking to Humboldt about the difficulty
of the question concerning the Mammoths and their preservation in Siberia, he more fully explained the views developed in his work. 'All these animals,' said he, 'are
found in foreign soil, and we know that men even have
been preserved in like manner. Thus, Prince Menschikoff,
who was exiled to Siberia and died at Obdorsk, was buried
there in full uniform, with all his medals and orders; and
on opening the ground a few years back, his Excellency was
found nicely preserved, moustachios, orders, and all, and
much more perfectly than if he had been embalmed. This
disinterment caused a good deal of hubbub, as it was done
without the authority of the Priest or Papa, and the ex-
cavators went so far as to pluck the Prince's moustachios
from his face and send them to Russia in proof of the fact.
Just as Prince Menschikoff has been preserved, so may the
Mammoths, who had wandered beyond their usual line of
travel, have fallen into crevices, and been potted up in
frozen earth.'

"Speaking of politics, I asked him how matters were
going on here, and if there was really much excitement.
He seemed to say it was much exaggerated. 'But,' said
he, 'the King reads everything, even your *Times*, although

it did lately say of him, that if he persevered in his present
measures, he would render himself " perfectly contemptible." '

"He told me he had made a curious hit in reading the
Timæus of Plato, as given by M. Martius. He finds in the
mouth of Polonius, the Jew, certain theoretical conjectures
respecting the gradual uprising of Continental masses, which
seem to harmonize exactly with the views of modern geologists. I told him of the King of the French's remark
about the 'growing of the land of Norway,' and the anecdote of the old Norwegian priest with whom His Majesty
conversed when a wanderer about the world.

"But the hour of three was approaching, and the Baron
got into 'double quick' to be in time for the royal table, and
I adjourned to the railroad 'restauration,' where I dined,
smoked my cigar, and wrote these memoranda." . . .

"I cannot leave Prussia without again confirming my
observations of former years. The troops of the line, cavalry
as well as infantry, are chiefly boys (I speak of the mass),
and the Landwehr are the best soldiers. How can cavalry
be worth anything, when a dragoon goes to his home after
three years' service, just when he is well formed? How
keep horses in condition with lads from the plough? How
have good gunners after two years' practice? Hence the
raw and awkward appearance of the sentinels, even at Berlin.
The system must be changed, or Prussia is sure to lose her
first campaigns against any old and well-disciplined army."

In entering Poland it was Murchison's intention to
gather from museums, professors, bergmeisters, and every
other available source of information, besides actual inspection of the rocks where visible, the nature and position of
the geological formations bordering the Russian tracts. At

Warsaw he fortunately secured the services of Professor
Zeuschner of Cracow, who accompanied him into the Carpathians, and whose previous knowledge of these regions
helped to save time and to make the tour more useful and
instructive than it would otherwise have been. Among the
endless geological details of his letters and note-book, there
occur occasional entries which show that the rocks were not
the only attraction in Poland.

"Here I am in Warsaw still. *Le beau ciel d'Italie* was
followed by storms which have been so violent that I am
not sorry to have delayed my departure a little. To tell the
truth, I wished to see the Mazurka well danced, and as the
devil and bad weather have willed it, the best dancers have
been ill, and the Colonel directing the ballet has not been
able to gratify me yet.

" Whatever changes come over Poland, her charming
women will never change. May they always preserve their
sweet manners, warm hearts, and generous sentiments ; with
such qualities they will improve the race by whom they
have been conquered."

Received everywhere with a frank and hearty hospitality
which charmed him, Murchison seems to have exerted his
utmost to please his entertainers. At one country-house
we find him recounting the pleasures and perils of the
Ural Mountains, and the march of the Siberian exiles, to an
audience to whom anything about Russia or the Russians
had an absorbing interest. Again, he is in the midst of
Polish national songs and dances, making minute inquiries,
and showing the keenest personal interest in the charac-
teristics of the conquered and partitioned kingdom. At
another time he keeps a family circle amazed by stories of

English railways, tunnels through mountains, and a scheme for making a roadway between France and England under the Straits of Dover. And thus, even where interminable sand and boulders concealed the rocks below, and deprived the geologist of one great source of pleasure, he made up for the loss by many a pleasant hour in the midst of the inner domestic life of Poland.

Getting out of the plains into the valleys and ravines of the Tatra range of the Carpathians, he and his companion had sometimes to wade knee-deep in snow. They made many traverses of the rocks with the view of comparing the structure of the country with that of the Ural chain. Amid the heaps of detritus in some of the valleys, he speculates on the former presence of glaciers, but regards the grand source of all the gravel and waste as traceable not to any superficial action, but to the upheaval of the solid nucleus of crystalline rocks through the secondary formations at the time of the birth of the Carpathians! In such observations as these we see how completely the early lessons of waste, taught him by the valleys of Auvergne, had been forgotten, and how thoroughly he had identified himself with the cataclysmic school of geologists.

Returning from the Carpathians by Cracow to Breslau, Murchison turned aside to make a section through the chain of the Riesen, Erlitz, and Sudeten Gebirge, by Freiburg, Waldenberg, and Glatz into Bohemia. Getting through the hills, he found himself on the interminable plain of northern Bohemia with its fortresses, stopped at one gate by the challenge, " Sind Sie Baron oder Graf? " at another, by being carelessly driven against a wall, and at last brought to a stand by the complete collapse of his broken carriage,

which had to undergo repair "in the most stupid of all
little towns, without a stone or a quarry near it, and in
the very middle of a great plain, the base of which is
Pläner-kalk, and the covering gravel and mud. What a
punishment on this earth!"

At Prague he met a man with whom he was destined to
have in future years much intercourse and correspondence,
the illustrious Joachim Barrande. "This very remarkable
man," so he wrote at a much later time, "was the tutor of
the Duc de Bordeaux, and was selected for that office from
the *École Polytechnique.* When Charles x. abdicated, Barrande, being attached to his young pupil, accompanied the
ex-Count to Prague, and soon being undermined by the
parti prêtre, he gave himself up to natural history studies.
In a trip to Vienna he first saw my 'Silurian System,' and
at once recognising the great similarity of the Bohemian
fossils to my own types, copies of which he made with his
own pencil, he from that day went to work steadily, found
and described hundreds of new forms, and finally made one
of the most classic works of our age, the *Système Silurien
de Bohème.*

"I have had of course long and continued intercourse
with this gifted and excellent man for the last twenty-two
years (I write this in 1865), and every year I have learned
to admire and esteem him more and more."

After some time spent among the Silurian rocks of the
Prague basin, and arguments with Barrande about them and
their fossils, the traveller turned north again into Saxony.
At Dresden art and art-criticism for a few days took the
place of the science which had for so many years driven
them out of Murchison's note-books. On the 19th July he

re-appeared in Berlin.   Under this date the following entry occurs in his journal :—" This is a proud day for me.   A budget of letters awaiting me from Warsaw, besides most agreeable letters from Tcheffkine, Helmersen, and others, contained one from Count Cancrine, officially announcing to me the transmission of a monumental present of the Emperor for all my services.[1]   The inscription on the porphyry pedestal is :—

GRATIA IMPERATORIS TOTIUS ROSSIÆ

RODERICO MURCHISON

GEOLOGIÆ ROSSIÆ EXPLORATORI

MDCCCXLIII.

whilst the steel plate on which the colossal vase stands, damasked at Stataoust, has on it in Russ :—

'TO THE GEOLOGIST MURCHISON

IN TESTIMONY OF ITS PARTICULAR ESTEEM.

THE ADMINISTRATION OF MINES

OF RUSSIA.'

How shall I ever render my work worthy of such a largess ! So now to bed to sleep over my honours."

Official rules debarred Murchison as a British subject from wearing foreign orders in Britain.   Efforts had been made to obtain a relaxation of these rules in his favour, even the philosopher and courtier Humboldt interesting himself in the matter.   The arrival of these fresh tokens

---

[1] This was the great vase of Siberian aventurine, four feet high, and six feet in circumference, which henceforward formed one of the most prominent objects in No. 16 Belgrave Square.   It was bequeathed to the Jermyn Street Museum, where it now stands, with its massive porphyry pedestal.   Owing to the difficulty of obtaining so large a block, and of polishing such a hard material, only one other similar vase has been made, viz., that presented to Humboldt, and now in the Royal Museum, Berlin.—See Bristow's *Glossary of Mineralogy*, *sub voc.* Aventurine.

of the esteem in which science and scientific men were
held in Russia, raised the question again as to the orders
and the general treatment of men of science by the British
government. This subject formed one of the topics now
gravely discussed by the geologist and the illustrious traveller at Berlin. We return to the journal.

" 21*st* *July.*—A royal day, devoted to the King of
Prussia and Humboldt. I went to Potsdam by the railroad,
and saw the great traveller walking with the crowd from
the station. . . . We went to work upon the Carpathians,
several results of my tour in Russia, Count Woronzow, and
many topics.

"I read him the letter I had had from Helmersen, in
which, *inter alia*, he speaks of the *ouvrage sublime* of M.
Humboldt. His eyes brightened at this unexpected praise,
and he said it was the first kind word he had had from
Russia concerning his last work. A snuff-box, indeed, in
diamonds, with a portrait of the Emperor, he had received ;
but these and official documents he valued slightly in comparison with such unbought praise.

" On producing my documents from Cancrine concerning
the vase and the Emperor's kindness, he at once said, ' This
must be made known to the public in justice to all men of
science, and to prove how they are appreciated in Russia.
Besides,' said he, ' after the unpleasant circumstances attending your decoration in England, I should like to let your
ministers feel a little.' So taking the documents and inscription, he added, ' I will see that this is noted with a slight
comment in the *Preussische Staats-Zeitung*, and then I hope
it may find its way into your papers. But if not, you ought,
in justice to the Emperor, to have a notice of it inserted in

any paper which M. Peel reads.' He assured me he had made every effort with the great Sir Robert to induce him to relax the order in council respecting my foreign order, and again repeated to me what indeed he told me in England, that neither the Premier nor any of the leading persons seemed to have the slightest idea of the relative value of scientific merit. He has evidently the opinion that Peel is not a truly great man, but one who shrinks from noble efforts, unless interest or expediency leads him. Hence we went into discussions on various proofs of this aloof from my small concerns.

" We were in the midst of such chat, when his chasseur came back, saying, ' The King has had the letter, and the Englishman is to dine at Sans Souci.' "

The gossip of the journal regarding Humboldt is here interrupted by a full narrative of a Court dinner at Potsdam, where in a quiet unostentatious way the Royal Family received their guests, and where Murchison appears to have been vastly pleased. It resumes as follows :—" Travelling back to my inn and unbuttoning, I returned in plain clothes to the Baron's rooms at Sans Souci (for he has them in both palaces), and there we renewed our chat. He told me he had taken the liberty of writing to Peel *in re* Robert Brown, suggesting a pension for the *Princeps Botanicorum*, and stating that he did this entirely without Brown's knowledge. I, of course, lauded the effort as it deserved (indeed I had previously spoken of it to Humboldt, and Buckland had written), and added, ' I was sure that on this occasion his voice would prevail.' Still he seemed to doubt, and placed little reliance on what he called the *buckram* minister—the man of sees, and saws, and appliances.

" We parted at the great gate of the Royal Gardens, and I got back to my inn, packed up and rolled back to Berlin, having in company the King's architect, who had been to his Majesty with plans in the evening. As soon as these were disposed of, the royal party would assemble ' *en petit comité.*' That evening my friend Baron von Orlich was to show his Hindustani drawings; on other occasions Humboldt and others read new works and criticised them. Thus quietly and unostentatiously, happily and sensibly live the King and Queen of Prussia. Long may they so live, and God bless them!

" I was again in the Hôtel de Russie at 10 o'clock; wrote letters; slept five hours, and was up at 5 o'clock. Off at 7 in the Eisenbahn for Leipzig, and have written my day's work just as we reach Wittenberg, at 10 o'clock."

There were still some geological sections to explore in the Saxon duchies before Murchison could rejoin his wife. So he once more turned south to Leipzig, and then south-west-wards by Gotha and Eisenach to Berka. Under date 25th July the following entry occurs in his journal:—" Well may a geologist say he never can bespeak his bed! I had taken leave of the essentials of my work on the slopes of the Thüringerwald, and was bowling along at a merry pace from Marksohl to Varta, when, having nothing to look at but the so-called tiresome ' Bunter Sandstein,' I took up my memorandum-book of two months old, and found ' Reichelsdorf and surrounding country to see on my return,'—a note I had taken from Germar at Halle. Alas! I had passed by my game, and, after travelling two and a half German miles, was now as much from my point. Arrived at Varta, I balanced for a few seconds. On the forward and homeward side of the argument lay my wife, anxiously expecting me at Baden, the meeting of the British Association at Cork awaiting me, and the desire to reach a good inn at Frankfort without more bivouacking. Then again there was the bother of returning, getting into the Hessian bad roads, and being bored to death, and, after all, perhaps, to see little or nothing to repay me in the way of analogies to my Permian, amid the Zechstein, Kupfer-Schiefer, Grauliegende, and the overlying Bunter, with which the tract I was to explore is beset. A faint heart, however, thought I, would never have obtained the Emperor's vase, and ' Zurück nach Berka' was the word given, and a fresh pair of horses was at once harnessed to take me back."

At the Reichelsdorf mines he came upon a curious and credulous bergmeister, on whom the miners had played so many tricks by supplying him with fossil wonders of their own device, that his collection of sea-devils and all manner of unknown monsters had become one of the lions of the district. Our traveller writes, "I paid him eleven dollars for his whole collection! The fossilized spitz-hound or his own dog was, however, too strong even for him. 'What is this large flag?' said I, seeing one covered over in the drawer. 'Das ist nichts.' 'Aber lassen sie mich sehen.' 'Ich bitte, das thut nichts.' This increased my curiosity, and pulling out the flagstone I saw on turning it over, the dog ' Ulick'—fossil, all pyritized by the workmen."

It was the 29th of July before Murchison rejoined his wife at Baden. After various excursions and visits to geologists on the way home, they reached England in time for the meeting of the British Association, which had this year been fixed to take place at Cork.

There had been considerable political excitement during
the summer in the south of Ireland. So serious indeed did
the prospect appear, as reported in the papers which found
their way into Galicia, that Murchison wrote thence to
Phillips, gravely proposing whether it would not be wise to
ask Government if the meeting of the Association could
with safety be held at Cork. Such fears were enough of
themselves to make the success of the meeting at least
doubtful. But other causes stood in the way. It had been
originally agreed that York should be the place of meeting,
—a decision overturned in the end by a majority of the
General Committee. Those who favoured the Irish town
seemed to forget afterwards that considerable exertions
were needed to insure a good attendance, and to make all
the machinery of the Association work smoothly and harmoniously. When, however, the executive and the leading
members reached Cork, they found that no adequate preparations had been made, and that the visitors from a
distance were few in number. "That which we hoped
would prove to be a south of Ireland meeting, turned out to
be a mere city of Cork concern. It was desperately up-
hill work, and the few of us who had any position were
obliged to swell ourselves out and speechify, and jollify, and
make the best of a very untoward thing. We were never so
near shipwreck as at this Cork meeting. For myself, I was
so imbued with our desire to go to York that I was constantly putting in that word instead of Cork."

It will be remembered that in the early discussions
regarding the order of the rocks in Devonshire, reference
was now and then made to the rocks in the south and south-
west of Ireland, whither, indeed, Murchison would have

betaken himself had he not been withdrawn by the greater
attractions of the Rhineland geology. Having now, however, got as far as Cork—a place he had not seen since that
memorable day when Sir Arthur Wellesley's expedition
sailed for Portugal—he determined to make a dash at the
older rocks which form the noble iron-bound coast line from
Bantry Bay to the mouth of the Shannon.[1] Several weeks
were given to this work, and at last after so prolonged an
absence, Murchison returned to his desk in Belgrave Square,
and the elaboration of the text and map of the work on
Russia. We get a pleasant picture of him and of his regard
for his old friend at Cambridge, from the following letter :—

" UP PARK, PETERSFIELD, *Oct.* 22, 1843.

"MY DEAR SEDGWICK,—Your shot from Dent sounded in
my ears in due time (and very agreeably) in Belgrave Square,
and I intended to write to you whilst you were in the north,
but De Verneuil was with me, and I was very busy about
Russia, and put off doing so till I went to Highfield and
this place, so I fear I have not let you know what all my
other friends in England know, that I am as large as ever,
and the picture of health after my Carpathian—Hibernian
tours. Why, where the de'il can you have been in your

---

[1] A souvenir of this journey may be given here. One of his travelling
companions, a clever and merry Irish girl, who beguiled the journey in
the mail by teaching him a small vocabulary of Irish, had remonstrated
with him as to the high price (two guineas) then charged for Ladies'
tickets at the British Association meetings. In parting he gave her his
own platform ticket, with the promise that if she should ever in after life
meet him, he would endeavour to befriend her. Years afterwards that
same ticket was put into his hands at the Association meeting at Birmingham. The girl was now a mother and a widow, in great difficulties,
and striving to gain a living by teaching. Murchison took immediate
steps to relieve her present wants, and to interest others in providing for
her future employment and comfort.

Cambria, that you never heard of the *Cork* and *York* meeting?

" I have loads of things to tell you, and if you say that you will positively be at the first meeting, *i.e.* 1st November, I will be in town to meet you.

" I have effected a great reform in the south of Ireland, in which there is not one bit of rock older than Devonian, excepting Ferriter's Cove, where the strata are all overturned. But have I not told you all this?

" Did I not also tell you that Sharpe has been hard at work (according to his own account) in Cambria, and will doubtless fire off *instanter*, as he was all full of it when I saw him at the Athenæum. He knew nothing of your having been in Wales, nor did I know of your doings at that time.

" I suppose your loyalty has brought you up to the Cam; if so, my reverence to the very illustrious Vice-Chancellor, who must now be the ' *beatus ille vir.*'

" If you *are* there, just fire off ten of your descriptive lines *after the show*, to amuse my wife and self, and direct them hither.

" Blessings on you (if they are worth anything) from your old smoking chum, and in foul or fair weather, believe me ever thine,   ROD. I. MURCHISON.

" *P.S.*—I am most anxious to show you my colossal vase from the Emperor of Russia, all the way from Kolyvan on the frontier of China—of Siberian Aventurine—weighing with the porphyry pedestal, a matter of two tons. So much for Imperial gratitude, albeit a man of science may work his hands off here and never be noticed by his Sovereign.

" R. I. M."

The epithet "smoking" applied to himself by the writer
in this letter was an appropriate and distinctive designation.
The habit which, as we have seen, marked him out in old
days in the hunting-field, remained as strong as ever. So
great indeed was his love of a cigar or a pipe, that in the
winter of 1840-41 he set on foot a movement in London to
found a new club, to be called "The Smokers," or "The
Raleigh," the fundamental basis of which should be the free
use of tobacco. He himself framed a prospectus of the
undertaking, wherein he drew a gloomy picture of the
miseries of smokers in the existing clubs of London, either
precluded from the solace of a "whiff," or, if permitted to
indulge their taste, banished to some scanty and cheerless
attic. In the new fraternity everything was to be designed
with a view to insure the most untrammelled enjoyment of the
weed. Special importation of the best tobacco, well stocked
wardrobes for the convenience of such members as might
have occasion to quit the atmosphere of smoke for social
intercourse with the outer world, good wines, artistic cook-
ery, and suitable literature were among the attractions of
the prospectus. This document, sent by its author to J. G.
Lockhart, drew from that caustic friend a brief and characteristic note :—" Your Grace's puff is quite admirable ; it
could not be mended were Raleigh to rise for the purpose.[1]

---

[1] It was a favourite and almost inveterate joke with some of Murchison's friends to quiz his love of rank and position by styling him "Duke
or King of Siluria," "Lord Grauwacke," or some other title referring
to his scientific work. Lockhart, who often sent short notes to him,
usually addressed him as "Your Grace," or "Your Highness," and after the
Russian campaigns as "Dear *Grand* Duke." At the same time Conybeare congratulates him on his Muscovite successes, beginning with "Dear
and most illustrious Count Silurowski Ouralowski." Murchison rather
liked this sort of thing.

I wholly decline the honour of belonging to the club. I
have no club habits, and hate especially all smoking in a
room but what is solitary."

The winter of 1843-44 found Murchison full of work over
his Russian volumes, correcting proofs, and carrying on a large
correspondence with friends in this country and on the Continent as to the rocks and fossils which he had to describe,
but escaping, as of old, for a few days' shooting now and then
at Up Park or elsewhere. One of the questions which occasioned a good deal of flutter in the scientific ranks during this
season was the determined opposition shown by Whewell to
certain proposals of the leaders of the British Association.
There had been a pretty general feeling that in its cycle of
perambulation that body should begin again with the towns
in which it had held its earliest meetings. We have seen that
York had been almost fixed upon for the assembly in 1843.
That town had now been selected for 1844, and if the former
order were to be observed, Cambridge would entertain the
Association in the following year. Whewell, however, set
his face most persistently against this proposal. To conciliate him, the Council proposed to choose some other place
for 1845, and to take Cambridge next in order. But he
declared that this would be equally objectionable, grounding
his argument on the law and practice of the Association in
favour of a wide range of places to be visited. Even in
Cambridge, however, his friends, such as the Dean of Ely,
and Sedgwick, refused to support him, and energetically lent
their assistance to the Council of the Association. Murchison, of course, had his full share of meetings and letter-
writing on the subject. From his letters the following may
be selected as having still some interest, inasmuch as it well

defines the position which, according to one of its founders, the British Association should aspire to fill :—

"BELGRAVE SQUARE, *March* 1, 1844.

"MY DEAR MASTER,—As you have written to me with perfect candour concerning the future meetings of the B. A., you will, I know, permit me to reply to you in the same strain. We had a meeting of the Council yesterday, and I communicated, as you desired, your sentiments, etc. You must not be surprised when I tell you that the opinion of all present (and I know it to be the general feeling of the B. A.) is opposed to your own concerning the regulating principle of the Association. We repudiate the idea that the chief aim of our existence is to stir up a few embers of latent scientific warmth *in the provinces.* If, indeed, that were truly our *main* object, I for one would cease to play pantaloon or clown in the strolling company, even if it should have a benefit night, as you suggest, for the followers of Caractacus on the frontiers of Siluria ! We think that nearly all the places you enumerate are wholly incapable of receiving the B. A. in its present stature, and if it is to pine away in size (as at Cork), the body can no longer enact the part which entitles it to the *nation's* confidence.

" In such case it could no longer be what *it has been,* a parliament of science, which finds the *ways and means* of carrying out researches which, without its stimulus, would never be undertaken ; nor could it, with such poor backing as Portsmouth, Shrewsbury, etc., pretend for one moment to act by public opinion upon the *Government* of this country. Unless we have full meetings, our funds fail, and we can no longer *institute the first experiments* which, satisfying public men of their usefulness, lead them to adopt our recommendations. This very year the Government have taken up works *begun* by us to an extent of £1500.

" But how are we to get the guineas—how raise the ' rint ' if not supported by the strong voice of the *real science* of England ? It is not enough to go about with a begging-box *if our O'Connells leave us.* Now, admitting with you that it is by no means necessary or even desirable that the Association wheel should go the same round, catching up its old friends (none of them we hope *off work*), *nolentes volentes,* still it is essential to our well-being, if not to our existence, that we should now and then secure the embraces of a university ;—it is, I say, indispensable to have from time to time a fresh infusion of scientific blood, and a rally of our oldest and best friends, etc. ; if so, where (Oxford being lost in her tracts), I say, can we obtain such except in your honoured *Alma Mater ?*

" But whilst I first argue my case *con amore,* I at once admit that the very *look* of the Master of Trinity when he chides his foster-child, is entitled to the greatest respect ; and I for one can imagine no good and effective meeting of science at Cambridge in which he does not co-operate. I know and have known his strong objections to an early meeting there, but I venture to hope that to oblige *all his scientific* friends from whom he differs on *this* point, he will so far relent as to allow us to revisit a place so dear to us, at no very distant day. My proposal, therefore, is, or rather my urgent request is, that after chastening us and compelling us to break our cabalistic cycle, by making us take a new place for our first meeting after York, our good Master will again receive us with open arms, and that however repugnant a meeting might be to him in 1845, his opposition having sent us to fresh pastures in that year, he will once more put us into condition (pardon my old habits) by a Cambridge training. Now, as clerk of the course which the B. A. has to run, I have got the fresh pastures ready for 1845, wherein we may fatten. In the name of the Corporation, inhabitants, and science (such as it is with Phil. Duncan at its head), Bath has invited us warmly to visit her in 1845, and with Lord Lansdowne as a President, we should then have a good show and collect a good purse, for it is the centre of a net-work of railroads open to Ireland and the south-west, and four hours from London. But good as it may be, the Bath meeting would sound as a great *bathos* in our prospectus, if followed by the poor diets of *smaller* provincial towns, and then will come the very nick of time, when Cambridge can *reinvigorate* us. Pray, therefore, unite with our scientific friends at Cambridge, who are, I hear, far from being confined to the Dean of Ely and Sedgwick, and by allowing us to announce that Cambridge will succeed to Bath, assure the public that we still possess within us the *national scientific strength.*"

It was ultimately arranged that the Cambridge meeting should be held in 1845, with Sir John Herschel as President.

Among the younger scientific Societies of London, the Geographical occupied at this time a very inconspicuous position. Founded in the year 1830 chiefly by members of the Raleigh Travellers' Club, it was designed to foster the progress of geographical research by collecting and publishing narratives of travel; by forming a good consulting library of geographical works, especially of maps and charts; by keeping illustrations of the best kinds of instruments for

exploration in different climates; by aiding with suggestion
and information any traveller about to explore; and by entering into correspondence with other geographical and scientific
societies, and with persons interested in geographical discovery in all parts of the world. At the original meetings
when the Society was organized, Murchison attended, and
he showed such interest in the Society's welfare that in
1843 he was chosen President. Its fourteen years of life,
though by no means without vigour, gave little promise of
the dimensions and importance which the Society subsequently attained. How this success was reached, and how
intimately Murchison was associated with it, will be referred
to in later chapters. For the present we see him in the
chair of the young and still struggling Society, reading to
them, in the early summer of the year 1844, the first of
those anniversary addresses for which in his later years he
was perhaps more widely known than even for his geological
achievements. Turning over the pages of that early address,
we see the germ of all those which succeeded it—a broadly
sketched outline of geographical progress over the globe, with
sagacious forecasts as to where explorations should be carried
on, and what ought to be looked for, and with a blending of geological exposition which gave a scientific meaning
and cohesion to scattered and unconnected observations.
The first of his addresses is stamped too with a feature
which marked all his discourses to the Geographical Society
—a painstaking analysis of the work of foreign travellers,
and a generous recognition of merit wherever it could be
found. Undoubtedly, this characteristic has done much to
give the Geographical Society of London a position of
weight abroad.

These duties at the Geographical Society, the revision of proof-sheets of the Russian work, and an unexpected visit of the Emperor of Russia to London in May, when, of course, the explorer of the Ural Mountains gladly renewed his experience of that monarch's courteous and even friendly bearing to him, kept Murchison longer in town this year than usual. It was the beginning of July before he was ready to start, and as he had to be back in September to be in time for the British Association meeting at York, he had comparatively little scope for an extensive tour. There still remained one great conterminous region to be visited for the completion of the Russian map. The Scandinavian peninsula had already yielded an abundant series of Silurian fossils, and Murchison had often quoted them, but he had never seen the country from which they came. His plan for visiting that part of Europe is thus told by himself :—

" MY DEAR PROFESSOR FORCHHAMMER,—I have resolved to visit Christiania at the meeting of the Scandinavian philosophers under Hansteen and his conjoint Presidents. I was for some time undecided about it, as I wished to get my great work on Russia finished, but finding that this is impossible before the early part of the winter on account of various delays which must always occur in extensive scientific publications, I have resolved to take flight for Hamburg and Copenhagen in the first days of July, and hope to find you still at home, that we may go on to Christiania together.

" My intention is further to traverse the country from Christiania to Stockholm, and to return by the isles of the Baltic. By this hasty visit and *your instructions* I hope to render my map, which embraces a good part of Sweden, somewhat more perfect. At all events I shall see the source of all my old friends, the erratic blocks, and look at some Silurian relations *in sitú.*"

Travelling rapidly into Denmark he halted for some days at Copenhagen, paid his respects there to King Christian VIII., whom he found to have some knowledge of geology, met the scientific men of the city, and revived his love of art-criticism among the bas-reliefs of Thorwaldsen. Arriving in time for the meeting of the savans at Christiania, he found it a very different affair from his British Association. "The scientific meeting," to quote from his journal, "opened by a general assembly on the evening of our arrival, when the first president, Hansteen, sat still, and the third president, Dr. Holst, announced in a solemn manner the laws and method of election, and the distribution of time. The hours of meeting were so managed that no two sections were ever sitting at the same time; and as no section works more than two hours, every man may cull from any school he pleases. Besides these, there are three general meetings in the hall of the Storthing.

"At our first meeting the medical men were desired to leave the room, and in a trice three-fourths of the chamber were seen moving into an adjoining room to elect their presidents and secretaries. The plural number must be employed, for the apothecaries have a separate medical section distinct from that of the doctors! The spreaders of plaster and lint and compounders of medicines play an important part in Scandinavia, and are as necessary to the doctor as the attorney to the barrister in England. The migration of all those sons of Esculapius at once showed me the stuff of which the mass of the meeting was composed. Out of the remaining fourth we had to form all the other four sections.

" Grouping together in a corner of the hall, we geologists chose our own President and Secretary. Forchhammer addressed us in Norse, saying, as I was told, that Von Buch declined to take the chair because I was here, and that I was *princeps inter geologos:* he therefore proposed me as President! To this I warmly objected; not wishing to be a King Log, and having no knowledge whatever of the language, I urged that I should be in a false position. Notwithstanding, however, all I could say, I was all but elected. But, to my great delight, and with my vote for him, M. de Buch was chosen in spite of himself. He is really a wonderful man, since he talks Norse very tolerably, and is quite able to explain himself on all points.

" At the first general meeting held on the second day, from one to half-past three o'clock, we had three subjects : 1*st*, Hansteen ; 2*d*, Oersted, or the identity of Electricity and Magnetism ; 3*d*, Retzius, on the Races of the Human Species.

" This done, we all huddled away to the Freemasons' Hall, a great pink building at the south end of the town, where all the philosophers and their wives, and also the Ministers of State, and a bishop or two, with the consuls of France, England, and Russia, were assembled, first in a great room above stairs, and afterwards in the dinner-hall beneath, where we sat down, nearly two hundred in number, at three tables. We had a bad dinner of four dishes, but it was very agreeable, and conducted without a trace of confusion.

" In the evening the geologists went with Keilhau (Von Buch and all of us following) to an upper part of the valley

north of the town, and 200 feet above the sea, to look at
serpulæ of existing species, *in sitû* upon the surface of the
rock.

" *Thursday Morning.*—Here I am at my last day, and
have scarcely had time to say a word of the meeting.
Though up at six o'clock, the day has usually been consumed, on my part, in eating, drinking, talking, and twice
giving lectures of an hour and a half long, till the evening
has arrived, and then *fêtes* go on till midnight, with no darkness. The table of meetings, etc., will best explain how the
different affairs proceeded. Hansteen made a quiet President; but neither he, nor his second, Boeck, nor his third, Dr.
Holst, possesses any eloquence for the social part of the concern. It was in consequence of this that my little tirades
at the tents of the students in the Botanical Gardens, and
again at the last public dinner on Thursday, produced quite
an enthusiasm. Hansteen literally read his after-dinner
speech in proposing the King's health !

" On the last evening they had toasted the King of
Sweden and Denmark, and had got low down in the lists of
toasts, connecting my name with the British Association for
the Advancement of Science, when, in reply, I took leave
as a royalist, to propose the health of the Stadtholder, the
representative of his sovereign, and a type of Norwegian
hospitality; for although his excellency had fed us well,
and was seated at our table, the philosophers had quite forgotten him. The old Count Löwenskiöld replied in a brief
but very energetic speech, and seemed much gratified.

" But my previous ' let off,' when my own health was
connected with old England, was the most telling, because I
coupled it with my delight at seeing science more honoured

in Norway than in any kingdom on the earth. This is quite true, and not merely complimentary."

Before leaving the Scandinavian assembly, Murchison received from his friend Élie de Beaumont the gratifying intelligence that on the 1st July he had been elected correspondent of the Institute of France. There were five candidates, and their position at the voting was as follows :—

Murchison, ...	27
Fournet, ...	7
De Charpentier, ...	3
Sedgwick, ...	3
Freiesleben, ...	1
	41

To see something more of Scandinavian geology and geologists, Murchison passed on from Christiania to Stockholm, visiting Gottenburg on the way, and making the acquaintance there of the future distinguished Professor Lovén, who accompanied him during most of his stay in Sweden, and with whom he soon came to be on the friendliest terms. At Stockholm we find him once more at Court and full of enthusiasm over the kindliness and courtesy of the royal family. He spends his evenings sometimes with Berzelius, at whose house he meets with the best scientific society of the capital, sometimes hearing a lecture from Retzius, or sipping a glass of punch with Lovén. The days are given to geology, and specially to the marvellous examples in that region of the striated and polished rocks which had now been universally recognised as in some way the work of ice. He is still full of the notion of submergence and icebergs, and chronicles with much wonder the size and extent of the huge ridges or *osar* which, like the gravel eskers of Ireland, and kames of Scotland, run over the country in the most striking and puzzling way.

It was part of his plan to revisit St. Petersburg to consult with Count von Keyserling as to their joint undertaking, and particularly to receive from that indefatigable explorer the latest new data for the geological map of Russia. Another object not so prominently put forward, but doubtless having its full influence in drawing him once more to the banks of the Neva, was the presentation of a gold medal to the Emperor Nicholas, which had been struck in honour of that sovereign's recent visit to England, and which the geologist had undertaken to convey to his Majesty. With these incentives he accordingly turned eastward and up the Baltic, coasting the Aland Isles and the indented shores of Finland, landing at every halting-place of the steamer to make fresh observations regarding the rocks with their striæ and groovings, and finally getting to St. Petersburg on the 24th of August.

The only items of general interest which can be gleaned from the letters and journal of this time refer to the Emperor Nicholas. That monarch had recently been plunged into deep grief by the death of his beloved daughter the Grand Duchess Alexandrina. But he sent for Murchison, and talked freely with him on a great variety of topics. The journal gives us a very characteristic picture at the outset of this interview. Murchison had driven down into the country to the Peterhoff Palace, and was waiting in the antechamber, when he found, standing in one corner, his friend Sir William Allan, President of the Royal Scottish Academy, with his picture of Peter the Great. "Without my warm encouragement this excellent and modest man

would not have been here to-day. A week before, when I
went to see his work, he told me that he was on the point
of departure by the next packet, and that, owing to the dis-
tressed condition of the Imperial family, he had given up
all hopes of seeing the Emperor, though when in Scotland
as Grand Duke Nicholas, the Emperor had known him well,
had bought pictures of him, and had charged him never to
come to Russia without seeing him. Thereon I stimulated
him, saying that it would be a dereliction of his duty not to
announce himself, and I urged him to do as I had done, and
write to Count Orloff, and then leave the case to his
Majesty. If he did so, I offered to bet that he would succeed. The result proved that I was right."

While chatting in the waiting-room, Murchison, to his
consternation, discovered that the medal had disappeared!
Search was made everywhere, but in vain, and he had to
make his appearance before the Czar without the object
which it had been his ostensible mission to present personally. The stolen medal was afterwards recovered by
the police in a jeweller's shop, where it had been sold by
the driver of the drosky.

The conversation with Nicholas is thus reported in the
journal :—" The Emperor first asked me what people said
of the reason of his visit, to which I could only reply that I
was little versed in state affairs, and merely supposed that
he wished to see the progress we had made since he was
first in England. On which he interrupted me, saying, ' Ah
but ! no, no ! I had but one object in my journey, and
that was to study the personal character of your Queen.'
' Then I am sure,' I added, ' that your Majesty went away
well pleased.' ' Yes, indeed, I did,' said he, ' for a more

sincere and excellent woman I never conversed with. I
have the conviction' (these were his words in French) 'que
quiconque pourrait l'environner, elle n'oubliera jamais ses
anciens alliés, ni les véritables intérêts de son pays. I daresay,' he continued, ' you English think that your Sovereign
has so little to do with politics and state affairs that his
or her character signifies little, your ministry regulating
all things. That is very true as regards your home
affairs and colonies. But you will give me leave to be
of opinion that, as regards your relations to the Continent
of Europe, the personal character of the Sovereign of
England is of great importance to all of us who are your
natural allies.' He also spoke of other estimable qualities
of the Queen, and of how much good she did, of which no
public account was given. In short, he quite rejoiced
me by the tone and warmth with which he eulogized
Queen Victoria.

" ' There was a time, it is true, when I had some reason
to bear you a grudge ; things were going badly, but now all
is well settled, and I like the English in good faith. I am
yours entirely. My visit has made known to me the character of your Queen, of which I have formed the highest
opinion. I have every confidence in her, and though she is
bound to people who might draw her from the right path,
I am persuaded that she will never act against the true
interests of her people.' Here his Majesty no doubt alluded
to the Louis Philippe and Leopoldine influence, and the
danger of French and Belgian politics. ' My last letters,
however (continued he), gave me a good deal of uneasiness.
The situation is very difficult, and it is much to be feared
that you will be compromised in these African affairs !

God preserve us from a war, but if it breaks out,[1] I tell you, my good friend, as I have already told your Queen, that I have not a battalion, I have not a vessel that will not fight for you.'

"Then, alluding to the French Revolution of 1830, his Majesty went on, ' No, no, I don't change my opinions. A revolution such as that (1830) accomplished by a mass of villany and baseness could not lead to a fixed and stable state of things; I said so at the time, and shall always say so.'

"Returning to England proper, he said, ' I must own to you that with all my astonishment at your external progress, there was one sign which much grieved me : it was to see how much less respect the lower classes showed towards their Sovereign than at the time of my first visit; and I do not hesitate to attribute this marked change to the democratic movement which is passing at present through the whole of Europe.'

"On then asking after the health of the Empress, his countenance all at once changed, and taking my hand, which he warmly pressed, ' Yes,' said he, ' we have also had a terrible trial, but I have indeed had a test of the affection of my people which has touched my heart. As regards the death of my dear daughter' (and here the tears burst from his manly eyes), ' I wished to have her buried by night, without any pomp whatsoever, accompanied by a single battalion and some followers. What was then my astonishment to see the whole population from Tsarskoe Selo to the citadel (twenty versts) forming such a dense column that my horse could hardly walk through it. The sight of this multitude on their knees praying for us, and sharing in the

---

[1] We were then nearly at war with France.

deep silence the anguish of my soul, truly touched my heart. Then I felt what it was to possess the love of one's people.'

"The Emperor was near his dear child when she made her last sign to him to approach her, when kissing her and putting his ear to her expiring voice, she said, ' Papa, never forget the person to whom I owe everything.' This allusion was to Miss Higginbotham, the governess of this excellent and beautiful young princess, and no nobler trait of her character could she have left behind her.

"From Tsarskoe Selo to the citadel, the Emperor, at a foot-pace and on horseback, followed the hearse like a man smitten with so dire a grief, that he was unconscious of the world around. But when the pageant and the lights of the mausoleum were around him, and the imposing ceremony of depositing the remains commenced, after two or three convulsive efforts to restrain himself, he burst into a flood of tears, and for the first time during his reign, Nicholas was seen to be but mortal by his court and his faithful soldiers. There was not a dry eye in the church. The sternest grenadiers were in tears."

Quitting this painful subject, the Emperor gave his guest the opportunity of saying, in allusion to the falsehoods circulated as to his own character and that of his nation, that "it was only by using the pen and telling the whole truth about them, that these false reports could be suppressed in Europe." "Ah diable! as to the pen, I don't understand anything about it; that is not my trade. At the same time I am none the less grateful to my good friends who will tell the truth. That is all that I desire."

"I then (alluding to the origin and cause of the dislike to his government and person) ventured to speak out and tell him plainly that touching England the whole matter lay in the suppression of the Polish insurrection, and that if our public were once properly instructed as to the origin of that affair, and the total want of good faith exhibited by its chief actors, the false sentiment which had been raised by the Franco-Polish party would be extinguished. If, therefore, his Majesty approved of the whole subject being thoroughly canvassed and exposed, I knew that the editor of the *Quarterly Review* would like to get up a series of reviews on Poland. He replied, 'Do so, by all means, and say the truth. You know me well, and you know that is what I like. But with all your efforts you will have some trouble in showing the face of the cards. The Poles have suppressed everything. There are things which are only known to our family. For instance, to give you an idea of the origin of this insurrection, and of the manner in which these gentlemen behaved. There were suspicions at Warsaw that plots were being organized, and my brother Constantine wrote to me on the subject, and begged me to give him my opinion. I answered him at once, saying, " Lay aside all these suspicions; have full confidence in the Polish army. It is an army, of which the noble officers will never break their oath. The Polish army has never been unfaithful." Conceive then my surprise, when on the very evening of the day when my letter arrived, my brother being asleep without a guard, they tried to assassinate him !' The Emperor further said that if his letter to his brother had been published, he would have required no better defence in the eyes of every English gentleman. 'After the rising the insurgents took possession of all my brother's papers. He left them all his documents. Among these

there was a declaration written by his own hand that he
would never attach any credit to these kinds of conspiracies,
and he trusted entirely in the loyalty of the Polish people.
Why were these two letters not published ?   Why did
these Czartoryskis, and these people who had all my
brother's letters in their possession, suppress them ?   Why,
as men of common honesty, did they not let the truth
appear ?   Hence, I repeat, it is the truth only that I desire.
As for revolts of misguided people, they may be pardoned;
but the revolt of a whole army—an army to which I
confided the care of Poland—for I had not 10,000 Russian
troops in the kingdom, such baseness never can be forgiven, and as an old soldier, you must be of that opinion.'"

At the house of the Grand-Duchess Hélène, Murchison
records that "we had good fun with old quaint B—,
who is a capital specimen of a true German philosopher
and smoker.   When we left in the evening he had no
carriage whatever, having come in a street drosky, which he
had paid off.   Wishing to give him a lift, Von Keyserling
and myself picked him up on the bridge, and squeezed him
into our little calêche.   As soon, however, as we had driven
him about two versts, he exclaimed, ' Mais, j'ai un bout de
cigare dans ma poche qu'il me faut fumer.   Je ne puis
plus rester,' and pulling out the stinking half-smoked cigar
which he had in his pocket all the time at the table of the
Grand Duchess, he jumped from the carriage, and kneeling
on the ground, lighted it at one of the flaming pots of oil
which were burning in honour of the Emperor, and called
out for a drosky, in which he followed us to the Mineralogical Society."

On another occasion, at the same hospitable mansion,

Murchison obtained a promise from the Grand-Duke Michael
that a specimen of the great wild oxen of the Vistula should be
procured alive, and sent for the acceptance of the Zoological
Society of London.   There had been a good deal of merriment at the table, and the Grand-Duke, in taking leave of
the geologist, said to him, " Adieu donc, mon cher.
Rassurez-vous, nous vous enverrons un bison tout entier et
en l'ouvrant vous trouverez là-dedans votre médaille ! "

After this short and pleasant sojourn in St. Petersburg,
Murchison journeyed homeward by Berlin, where he
renewed his acquaintance with Humboldt and others, dined
once more with the royal family, and obtained still further
materials for the completion of his Russian map.   It was
in the latter half of September that he once more found
himself in England.

This northern tour had been unexpectedly successful in
its geological results.   To the flat unaltered Silurian rocks
of the Russian plains no base had yet been found, so
that neither Murchison nor his colleagues could tell what
the oldest fossiliferous strata of these tracts rested on.
But in Scandinavia he had found the old platform of
metamorphosed rocks on which the Silurian formations
reposed, and on which he believed the whole of the
fossiliferous deposits of the north of Europe must lie.
He had likewise learnt much as to the organic contents of
the older formations from the numerous museums to which,
through the friendly aid of Berzelius, Keilhau, and Lovén,
he had enjoyed unrestricted access.   The abundant illustrations of ice-action had likewise brought before him in a
new light many of the phenomena of that northern boulder-
drift which he had traced over the plains of Northern

Germany and Russia. Moreover, the conference with Von Keyserling at St. Petersburg bore fruit in the subsequently published volume on Russia, for that undaunted explorer had traced his geological lines through the wilds of North-Eastern Russia, a region in great part unknown, and stretching far beyond the limits of forest growth up to the shores of the icy sea.

In other respects the tour had been an advantageous one for Murchison. He had come into personal contact with the scientific men of Denmark, Norway, and Sweden, and formed acquaintances, and in some cases friendships, which, lasting through life, helped to broaden his hold upon the general scientific activity of his time.

Again, though the tour lasted little more than two months, it had brought the traveller into personal relations with four of the crowned heads of Europe. It was, of course, his steadily increasing reputation which secured him this attention. Though keenly alive to such social distinction, he was well aware that he owed it to his science, and he endeavoured to repay the obligation by losing no opportunity of doing his best to strengthen the position of science and scientific men with the powers that be.

Back again in London, he applied himself strenuously to the final completion of the Russian book, which had been so long in hand. So great had been the delay, so numerous the alterations and improvements occasioned by his own fresh observations in the adjoining countries, as well as those of Von Keyserling and others since the original surveys of 1840 and 1841 in Russia, that the estimated cost of the work was more than trebled. The printers' charge for corrections alone amounted to more than twice the sum originally destined for the printing of the whole volume. At last, however, in the month of April 1845, the preface was written, and in the course of the summer the first copies were issued.

Judged merely from the bibliographer's point of view, *The Geology of Russia in Europe and the Ural Mountains* was a magnificent work. It extended to two massive volumes, the conjoint labour of Murchison, Von Keyserling, and De Verneuil. The first volume was written by Murchison from his own notes and those of his colleagues ; the second, treating of fossils, was the work of De Verneuil. In the first volume, a *résumé* is given of what was known of palæozoic geology in Europe and America up to the date of publication. The successive geological formations of the great Russian plains are described, and then follows a detailed account of the structure of the Ural chain. The last chapters, among the most readable and generally interesting of the whole, treat of the more recent deposits, the formation of the Aralo-Caspian basin, the gravels and drifts so widely spread over the low grounds, the range of the Scandinavian boulders, and the extent and origin of the " black earth," and later alluvia. In short, the volume presented for the first time a clear outline of the geology of more than half of the Continent, and crowned the work which had been in progress ever since Murchison's earliest observations of the transition rocks on the banks of the Wye, by combining on one broad canvas a picture of the whole succession of the palæozoic rocks of Europe.

Though the task of putting this geological material together had been intrusted to Murchison, the honour of this great contribution to science was equally shared by his colleagues. They had taken part in the toils of the field, and contributed to the mass of detail out of which the general deductions had been evolved. The work was appropriately and gratefully dedicated to the Emperor Nicholas, who had taken throughout a lively personal interest in its prosecution and completion.

With the publication of *Russia and the Ural Mountains,* Murchison's position in the very front rank of geologists was universally acknowledged. Step by step, with unwearied industry, he had risen to this proud eminence. During the remaining years of his busy life he displayed the same untiring energy. But he had thenceforth no new kingdoms to conquer. His years, therefore, were passed in consolidating what he had reclaimed, and in witnessing the extension of his work by others into all parts of the globe. The story of this second half of his scientific life has now to be told.

SIR CHARLES LYELL, Bart., F.R.S.
From a Drawing by George Richmond, R.A.

CHAPTER XVIII.

THE first and most active half of Murchison's scientific career may be considered to end with the publication of the great work on Russia. Twenty years had passed away since he sold his fox-hounds and came up to London—an earnest and painstaking auditor at the lectures of Davy, Faraday, and Brande—twenty busy years, which at their beginning found him utterly ignorant of science and scientific methods, and at their close left him at the very head of the geologists of Europe. During that eventful period his life was one of constant activity both of body and mind. He had been late in starting as a man of science, but he more than made up for the delay.

We have traced how, after various essays in different directions, he finally settled down to the study of the oldest fossiliferous rocks as the business of his life; how, with great sagacity, he followed them from country to country, and made them reveal at last their contribution to the history of the earth and its inhabitants. It was not merely the geological structure of a part of Britain which he cleared up when he put these rocks into their true order and rela-

tions. In grouping them he discovered a series of early
pages in the story of the progress of animal life upon our
globe—portions of a record which, in its grand features, bore
evidence, not of any mere local significance, but of probably
a world-wide application. Following up his successes in
his own country, he confirmed the general application and
importance of his deductions by tracing this same great
geological succession over wide regions of the Continent.
His views had been adopted by geologists all over the
globe, and he had now, year after year, the gratification of
receiving, from even the most remote quarters, tidings of
the light which his classification threw upon the study of
the older rocks.

The main outlines of the scientific work of his life were
now filled in. Henceforth he busied himself completing
the details of the picture. To do this still required much
active research, which was devoted chiefly to various parts
of the Continent of Europe. But we trace from this time
forward a gradual relaxation of the engrossing hold which
geology had for these twenty years retained upon his
thoughts and affections. Not that he ceased to be as keenly
alive as ever to the progress of his favourite science, but with
the advance of years he gradually allowed geographical progress to claim a large share of his time and interest. He
identified himself heart and soul with the Royal Geographical Society, and for at least the last ten or twelve years
of his life was probably better known to ordinary newspaper
readers as a geographer than as a geologist. This was no
new current of thought to him. He had himself been a
traveller long before he became a geologist. As we have
seen, he was one of the members of the Raleigh Travellers'

Club who started the Geographical Society, and, while still
deeply immersed in the preparation of the work on Russia,
he had written an elaborate address for that Society, of
which he had been chosen President. In the remainder of
this record of his life, therefore, we shall find him not only
still wielding his hammer, but coming to the front as a most
earnest and influential promoter of geographical enterprise,
identifying himself even personally with the success and
the fate of travellers, and rousing the dormant sympathies
of his countrymen and government into active co-operation
with his own.

During the years immediately preceding the date at
which we are now arrived, much interest had been taken in
this country in the exploration of the Polar regions. The
first impulse to this interest had been given by the British
Association, which at the Newcastle meeting in 1838 had
adopted a series of resolutions strongly urging upon the
Government the propriety of fitting out an expedition to the
high latitudes of the southern hemisphere, for the purpose
primarily of carrying out important observations on the
subject of terrestrial magnetism. In 1839 Sir James Clark
Ross sailed with the "Erebus" and "Terror," and, after an
absence of four years, returned with a noble harvest of
results in many different branches of science. So brilliant
was his success that it rekindled the old passion for the discovery of the north-west passage, and thus, long before the
charming narrative of his voyage was published, the impatient nation had resolved that the "Erebus" and "Terror,"
which had so gallantly escaped all the dangers of the
antarctic ice, should not be allowed to lie idle in port, but
should be once more sent on a mission of Polar discovery.

In the month of May (1845), Franklin and his brave comrades sailed on their hapless voyage.

Among those who watched with keen interest the despatch of this expedition to the frozen north were the members of the Geographical Society of London. Their President, Murchison, in delivering to them his second and valedictory address in quitting the chair, spoke in hopeful words of the sailing of this new venture to discover a north-west passage. " Proud shall we geographers be," said he, " if our gallant Vice-President, Sir John Franklin, shall return after achieving such an exploit; and gladly, I am sure, would we then offer him our Presidential Chair as some slight recompence for his arduous labours." The interest thus expressed in the fate of Franklin remained active at the heart of the speaker through the long subsequent period of suspense and foreboding, stimulating him to earnest appeals for further inquiry as to the fate of the voyagers, until at last, after two-and-twenty weary years of uncertainty, the sad story was brought home by M'Clintock.

In resigning the Geographical Chair to his successor in office, Murchison referred with some satisfaction to the increase of members during his Presidency, and notably of " persons of high consideration." " He had striven," he said, " by every means in his power, to augment the members of the Society, to attach to its list names of distinguished men, and to render the Society as popular as it was scientific." The piles of letters still extant among his papers bear witness to the activity of his correspondence in this matter. During this his first tenure of the Presidency he began that system of recruiting and organizing which some years later he resumed when once more at the head of the Society, and to which the Society is confessedly in large measure indebted for its popularity and for the great assistance and encouragement which it has been able to afford to geographical research in all quarters of the globe.

The British Association assembled this year (1845) at Cambridge under the Presidency of Sir John Herschel, Sedgwick taking the chair among the geologists. Murchison contented himself with exhibiting an early copy of the *Russia and the Ural Mountains*, and contributing some communications which had been made to him from geological friends abroad. He had the gratification, however, to find both his devotion to the Association and his scientific reputation acknowledged by his election to be President at the meeting to be held in 1846 in Southampton.

It had been determined that the authors, who had dedicated their work to the Emperor Nicholas, should present it to him at St. Petersburg in person. Accordingly, Murchison and De Verneuil once more bent their steps together to the Russian capital, agreeing that after they had deposited their volumes there they should take a conjoint geological tour through the south of Sweden, where so much of interest had been noted in the previous year.

The earlier portion of Murchison's journal of this tour is devoted mainly to the sights of St. Petersburg, and to the dinners and receptions by the Imperial family, where he was welcomed with every mark of esteem. From the Emperor himself he received a measure of cordiality and even of confidence which astonished as much as it delighted him. The following excerpts from the journal will show the footing on which he stood :—

" Alluding to a great work which he was engaged in

reading when I entered, the Emperor said, 'You see me
poring over the last revision of our penal code, which has
never been rendered uniform, codified, or adapted to the
advancement of the times. We had an immense mass of
undigested and ill-consorted materials, and it is high time
to settle them all down in a spirit more tolerant and mild
than of old. To legislate for Russia as she is, is no easy
matter for those who do not thoroughly understand our people,
their dispositions and condition. What can be so absurd
as these continued tirades against us, and especially against
myself, because we do not regulate things as you do in
England or France ? As for myself, these gentlemen, the
journalists, detest me, because they look on me as the rock
which will never yield to their assaults. I hold myself
responsible to God alone,' added he, showing much intensity
of feeling ; ' it is for Him alone to judge me. It is He who
has placed me in this responsible position, and I will never
change in order to please (you will pardon me the phrase)
that *canaille*. No, I will never govern as a king of France
or a king of England; the respective conditions of our
peoples are entirely different, and what goes well with
you would lead us to ruin. It is I then, I who am the
rock against which these gentlemen bear a grudge. The
worst of all this crew are found from time to time among
our Russian renegades.'

  "I naturally again expressed my regret that His Majesty had not a few able penmen to defend him. But he
interrupted me saying, ' No, I do not vex myself about
these attacks; my conscience absolves me; I do not fear
death, and I know that my people love me. I repeat, I
am Emperor of Russia, and not King of the French. They

slander me now, but posterity, judging my position, will do
me justice !'

" He then went on to tell me how much he was personally
obliged to me, that he had real gratitude towards an English gentleman who espoused his cause on a principle of
probity and honour. After I had declared that nothing
would give me more sincere pleasure than to serve him in
extending science or in lowering his enemies, he took me by
the shoulders, and looking me through (for what an eye he
has !) exclaimed, ' I know it, my dear friend ; you are Murchison, you are an honourable man !' and then kissed me
first on the right cheek, and then on the left with a hearty
and real smack.

" I expressed to Count Orloff how much I had been
delighted with the Emperor's conversation with me, and
the expression of his attachment to England, Englishmen,
and our Queen, and congratulating myself upon the good
old alliance being sure to be maintained, I asked him if he
(Orloff) foresaw anything which might disturb this harmony
between our nations. At first he said no, but correcting
himself, he said, ' Yes, there is one event which has recently
transpired, which possibly may lead to trouble, but when I
mention it, you will laugh at me. You have just sent an
ambassador to Constantinople, who, though doubtless a
very able and a very upright man, is the only Englishman
with whom my master will hold no friendly communication.
So strongly does he dislike Sir Stratford Canning, that everything proceeding from him will be viewed in an irritable
temper by the Emperor. Now you know,' he added, ' the
relations of my country and your own require to be treated
with great delicacy, and all abstinence from any irritating

subjects of dispute. Hence you can scarcely imagine how
I deplore the mission of Sir Stratford Canning, for I fear
that it may lead to grave misunderstandings.'

"The next morning (Thursday), the eve of our departure,
Tcheffkine was with us early, and announced, as a secret,
that the day before I had been named to a Great Cross of
St. Stanislaus, and De Verneuil to a Second Cross of St.
Anne.

"At midnight I went to bed, and was speedily awakened
by a serjeant with a dozen of crosses on his breast, who
came in with candles, accompanied by Madame Wilson's
Gregorio (the Russian servant), and holding in his hand
two ponderous despatches containing the new orders for
De Verneuil and myself. At first I begged that the conference might be deferred till daylight, but the serjeant per-
sisted, and kept his post until I sat up in my night-cap
as a Knight Grand Cross of Stanislaus, and signed the book
of receipt, when I gave the envoy a dozen of silver roubles,
and the tilega was soon heard to rattle quickly from the
door, and Mrs. Wilson's was again left to quiet at two
o'clock A.M."

From the festivities of St. Petersburg Murchison and De
Verneuil turned once more to field geology. Sailing down
the Baltic and skirting the shores of Finland, they began an
examination of the ice-worn rocks and an argumentation
thereon, which appears to have taken up quite as much of
their time and thought as the structure and fossils of the
ancient rocks. Murchison stood out stoutly for icebergs,
and traced, or believed he could trace, their impress all
along the Finland coast and over the whole of the southern
tracts of Sweden. De Verneuil interposed on behalf of

glaciers. Thus the debate was renewed which had begun so vigorously when Agassiz startled geologists by demanding whole continents of ice to account for the ground and polished rocks of the northern hemisphere.

When the two travellers landed at Stockholm they each found everywhere fresh materials in support of his theory. They took a tour through the southern portion of Sweden, contriving, amid the fascination of the ice-work of that region to make sections in illustration of the Silurian deposits which proved of service in future illustrations of palæozoic geology in Europe. A ramble by Upsala, through Dalecar-

Succession of the Palæozoic Rocks in Norway—(*Russia in Europe*, p. 10.)

o Gneiss ; a Sandstones, Limestones, etc. ; b Pentamerus limestone ; c Coralline lime-stone (Wenlock) ; d Calcareous flagstones (Ludlow) ; e Old Red Sandstone ; p Rhombic porphyry ; t Other eruptive rocks.

lia, and to the mines of Dannemara, brought them back to Stockholm, where they found Berzelius, and the other scientific men of that metropolis, ready to welcome them, and where they were received at Court. Pursued by furious storms, they prolonged their tour into Gothland and then into the province of Scania, where, at Lund, they met the well-known archæologist, Nilsson, who gave them a disquisition on craniology. "Nilsson asks," so the journal records, "as the greatest favour, the cranium of a Highlander. I have promised him mine when I die. He looked at it with special affection as coming so near his Phœnician type."

At Copenhagen, on the homeward journey, Murchison renewed his intercourse with the Danish acquaintances whom he had met in the previous year. In the King he found an unpretending but interested student of natural history, who had formed an admirable series of cabinets, and who himself led him through it, "happier in a museum," so the journal remarks, "than in governing a kingdom." On reaching Paris, Murchison learned that his wife was ill at Tours, to which place he at once continued his journey. It was the old Roman fever which had prostrated her. Already convalescent when her husband arrived, she was soon able to be moved to Paris, where a stay of some weeks was made.

While at St. Petérsburg Murchison had been in communication with Count Nesselrode and other dignitaries there regarding a proposed appointment to some post in the Russian service, the apparent object being to secure such a title and position in that service as would obviate official objections at home to the use of the Russian orders. Rumours of this project had reached London, and of course soon grew into an exaggerated story, which in due time found its way to Murchison. Writing from Paris to Mr. John Murray, he remarks, " I cannot imagine what can have put it into any one's head (except that Lockhart is fond of a joke) that I had been named a general in the Russian service! At first I thought nothing of the slip of your pen, because it struck me that in a multitude of letters you might have had one for a general officer on your table; but, meeting with Mr. H. Ellis here, I found, to my very great annoyance and regret, that this nonsense had been a topic of talk in London! It never was in contemplation to make me a military man again; all that was designed on the part of my friends in Russia was to have accorded to me a scientific post in perfect harmony with my labours past, present, and future, under the title of ' Inspecteur des Explorations Géologiques de l'Empire,' and with the rank of privy councillor." Before he left Paris, however, he received intimation that the project had been carried out, though in a somewhat different form. He was appointed by the Emperor of Russia an " Effective Member of the Imperial Academy of Sciences, with all the rights, privileges, and rank attached to that office in the Imperial service."[1] Shortly afterwards he was gazetted in London as having the Queen's privilege to wear the Russian orders.

The sojourn at Paris formed a pleasant interlude between the field-work in Scandinavia and the ordeal of feasting which, unknown to himself, was then in store for Murchison in London. He presented himself for the first time to the Academy of Sciences since his election. He likewise appeared at the Geological Society of France, and being now brimful of ice-work from his recent Swedish and Baltic experiences, was put up by his friends there to give his opinion. For the battle of " Glaciers *versus* Icebergs " was then being fought out in Paris as well as elsewhere. Durocher claimed him as a friend of the forlorn iceberg ; Martins as a partisan of the glacier. He found that the contest grew warm in spite of the subject, and that the majority of the French, as might have been predicted from their greater proximity to the Alps and Pyrenees, were glacier men. Nothing daunted, however, he stuck to his débâcles and ice-floes. He remained indeed true to this his early creed for many

[1] " This nomination gave me," he says, " a position in the Russian service intermediate between that of a Colonel and that of a Major-General."

years afterwards, until the advance of opinion caused him at
last to concede a sort of divided empire to bergs and
glaciers.

Mrs. Murchison had recruited in Paris, and could be
moved to England just in time for Christmas. Once more
in London, and turning to the work that lay before him,
Murchison found two tasks awaiting him, one the narrative
of what De Verneuil and he had been doing among the
ice-work and Silurian rocks of the north, the other the
preparation of his address for the British Association. Fortunately for him neither of these labours was very pressing,
nor, as he performed them, very arduous. For he had
hardly got well settled down at home when he received
the following announcement, which had been communicated
to his old friend and instructor at Oxford :—

" WHITEHALL, *February* 5, 1846.

" MY DEAR DOCTOR BUCKLAND,—You will be glad to
hear that yesterday, on our own spontaneous idea of Mr.
Murchison's claims to a mark of favour from his own
Sovereign, Sir James Graham, with my entire concurrence,
wrote to the Queen, advising Her Majesty to confer the
honour of knighthood on Mr. Murchison at the first levée.

" The value of the distinction will be that it was un-
solicited and unprompted, and that it is intended as a
recognition by the Queen of Mr. Murchison's services in
the great cause of science and human knowledge.—Most
truly yours,   ROBERT PEEL."

The proposal, as gratifying as it was felicitously expressed and wholly unexpected, Murchison accepted, amid
the warm congratulations of his friends and associates. It

was the second instance of the kind which had recently
occurred (Sir Henry De la Beche's knighthood having been
the first) to show that the valuable geological work done in
this country and abroad by British geologists was not
wholly unknown and unrecognised by their own Government. One immediate effect of the accession of dignity
was to call out a vigorous display of a truly English mode
of showing sympathy and goodwill. Murchison was in-
undated with invitations to dinner, not only from his old
friends, but from a host of new ones. Not much continuous
mental work could therefore be expected from him for some
time after the 5th of February. But by the beginning of
April he had been able to prepare his account of the Scandinavian icebergs for the Geological Society, and two months
later he presented a memoir upon the Silurian rocks, in this
way relieving his well-kept journals of all their voluminous
details, which now passed into the *Quarterly Journal.*[1]

The meetings of this Society remained pleasant, often
amusing, and even exciting. The older heroes had not yet
departed, while a younger race, sometimes of great vigour,
had gathered round them. The " cuffing of opinions," which
had been an early trait among the geologists, still continued
to draw to their evening meetings a good many listeners
who had no special geological bent, but who enjoyed good
humour and wit. Leonard Horner filled at this time the
President's chair, contributing no scientific renown to the

---

[1] On the Superficial Detritus of Sweden, and on the probable causes
which have affected the surface of the Rocks in the Central and Southern
portions of that Kingdom.—*Quart. Journ. Geol. Soc.,* ii. 349.

On the Silurian and Associated Rocks in Dalecarlia, and on the succession from Lower to Upper Silurian in Smoland, Oland, and Gothland,
and in Scania.—*Op. cit.* iii. 1.

Society, but guiding its affairs with singular sagacity and foresight. Under his sway the ponderous quarto Trans-actions, which few but professed geologists ever thought of buying or reading, and which from their costliness could only appear at wide intervals, to the detriment alike of the science and of the authors of the papers, gave place to the present *Quarterly Journal,* a convenient and regularly pub-lished record of the communications made to the Society. During his Presidency, also, while younger palæontologists were rising into eminence by their work among the fossils of the earliest formations—Forbes, Morris, Davidson, Salter, M'Coy, and others—a graceful act of recognition came from the Society to one of the older school, who, in enfeebled health, had quitted its service to seek rest in retirement. The gentle and ever helpful Lonsdale received this year (1846) the Wollaston medal, the highest honour in the gift of the Society.

Among those who kept up the old spirit in the gather-ings at Somerset House, none came more welcomed than Sedgwick. For several years past he had appeared more frequently as an author of papers than he had done for some time previously; and a paper by Sedgwick was sure to bring a full meeting. His subject had been the slates and older rocks of North Wales and Cumberland, a subject peculiarly his own, but on which other writers had recently put forth different opinions. As these papers have an intimate relation to Murchison's position, let us turn for a few moments to what the Woodwardian Professor had been doing in regard to his former Welsh work, the true mean-ing and value of which had been perhaps somewhat obscured by the more recent observations of Bowman, Sharpe, and the Geological Survey, as well as by those of Murchison. He felt and expressed at the time, in letters to the latter friend, that it was absurd to theorize about the structure of a great region, such as that of North Wales, merely from one or two small sections, and this was what he thought had been done both by Bowman and Sharpe; still he felt that a renewed examination of the ground would be of advantage to his views, and accordingly he had spent a portion of the autumns of 1842 and 1843 in North Wales, in the hope, as he wrote to Murchison, that he might succeed in showing, by fossil evidence, that there existed in his older or Cambrian rocks a type of life different from that in Murchison's Silurian series. But, as we have seen,[1] he had failed in that part of his quest. He could not find, any more than other observers, a geological line of division between Cambria and Siluria. He succeeded, however, in establishing one or two important facts in palæozoic geology.

For the first time, so far as appears, he showed that the so-called Upper Silurian rocks of Murchison over wide tracts of country, do not graduate downwards, as the latter geologist had represented, into the Lower Silurian, but rest upon them in what is called an unconformable succession. In other words, he proved that after the Lower Silurian rocks had been deposited, a long period of time elapsed, during which they were upheaved and exposed to waste at the surface before the Upper Silurian formations were laid down upon them. He maintained also, that hardly any of the species which lived during the older period survived these great changes of physical geography, so as to re-appear in the Upper Silurian waters. Hence he now

[1] See vol. i. p. 382, *note.*

regarded the older fossiliferous rocks of Wales as divisible
into two great groups :—1*st*, A vast lower series embracing
the so-called Cambrian and Lower Silurian rocks, which
he believed to be geologically united, and to which he
gave the name of the Protozoic group; 2*d*, An upper
series resting over wide districts, upon the upturned edges
of the Lower, and embracing the Upper Silurian formations
of Murchison up to the base of the Old Red Sandstone.
While tracing out the boundaries between these groups, he
had given renewed attention to the relics of ancient volcanic
action in North Wales, and again clearly recognised that
some of the igneous masses there had been ejected contem-
poraneously with the deposition of the aqueous rocks among
which they lie, while others had been thrust subsequently
into cracks of the overlying formations.[1]

With these views as to the vague boundary-line between
Cambria and Siluria, Sedgwick does not seem for some time
to have had any solicitude as to the ultimate decision of the
question. He had made the rocks of North Wales so
thoroughly his own, by years of successful toil, that the idea
of relinquishing his hold upon them and regarding them as
forming part of the domain which had been rightfully conquered by another, had probably never been seriously in his
mind. So far from looking upon the final determination of
the boundary-line as likely to curtail his dominions, he
seems to have contemplated the fusion of his own empire
with more than half of Murchison's into one great whole,
leaving only the upper and minor part of Siluria to retain
its distinctive name. And when, on further examination, he
found additional proofs of the significance of the great dis-

---

[1] *Proc. Geol. Soc.*, iv. 212. Paper read by Sedgwick, 21st June 1843.

cordance between the Lower and Upper Silurian rocks which had been missed by Murchison, he felt justified in after years in drawing the line between Cambria and Siluria where, as it seemed to him, nature had drawn it by that break in the succession, claiming as Cambrian the whole of his own series and all the Lower Silurian rocks of his friend.[1]

That the Cambrian system should have an upward extension instead of the Silurian system having a downward one, had been surmised by at least one geologist besides Sedgwick. The idea was broached by Phillips in a note to Murchison, who, in alluding to it in his reply, remarked, "I cannot read the paragraph which follows in your note without being hurt by the *possibility* to which you allude of the Cambrian going up to the Wenlock Shale. As a matter of fact, unquestionably it does. But what is Cambrian? Why, as De la Beche has shown in South Wales, nothing but *Lower Silurian.* Now, bear in mind, that my two clear and distinct types, Upper and Lower Silurian, were proposed to geologists, and, being laid before them, they were asked to see how far the lower type would go down. This is repeatedly put to them in my book."

The point in dispute must seem to outsiders to have been but a petty one. After all, it was merely whether one name or another should be given to a certain series of rocks upon which both of the geologists had been simultaneously at work. Had Sedgwick continued his researches to their proper development, had he taken the trouble to have his collections of fossils examined and named instead of allowing them to remain for years in their packing-cases, had he,

---

[1] Introduction to *British Palæozoic Fossils*, 1855, p. x.

by fossil evidence as well as by physical structure, shown
the distinctiveness of his rocks and their passage into any
other known geological formation, he would have established a good title to all that he afterwards claimed. He
did not do this. All the while, however, Murchison, with
less brilliancy and power, but with more industry and perseverance, was toiling to effect it among what were at the
time supposed to be other rocks, but which proved in the
end to be in great measure the very same as those of Sedgwick. He succeeded in determining their order, and showed
their relations to the next succeeding geological deposits.
He maintained that he had discovered a distinct type of
life,—what he had termed the earliest or Protozoic type.
He had given it a name—Silurian ; and he naturally
refused to alter his classification until it could be shown
that he had included more zoological types than one, or
that some still earlier type had preceded what he had supposed to be the earliest.

The question, in whatever way it might be settled, in no
way affected the title of either claimant to the honour which
he had well earned. As yet, indeed, they had not become
disputants. It is remarkable that neither at this time, nor
for nearly a dozen of years later, did they come to open
rupture.[1] This postponement of the disagreement must be

---

[1] There seems to have been some danger of a rupture in the summer
of 1846. Sedgwick had written a paper upon the geology of Cumberland, in which, without avowedly attacking Murchison's classification of
the older rocks, he indirectly opposed it. He speaks of the erroneous sections, by which the position of the Llandeilo flags had been incorrectly
fixed, places these flags in the same general line with the Caradoc beds,
and proposes a threefold division of the older fossiliferous rocks, viz. :—
1st, Cambrian, including the oldest fossiliferous rocks ; 2d, Middle group of
Lower Silurian, including Llandeilo, Caradoc, and perhaps Wenlock ; 3d,
Upper group, or exclusively Upper Silurian. This paper, when published,

mainly attributed to Sedgwick. Though naturally sensitive and impatient of the interference of others in his geological domain (he had once angrily resented a supposed interference of Murchison), he somehow failed to perceive, or at least openly to oppose, what must needs be the issue of his friend's strong contention as to the absence of any zoological significance for the so-called Cambrian series.

brought from the author of the Silurian System a letter to Sedgwick, in which the following sentences occur :—" I did not read your memoir until about a month ago, when my attention was called to it both by letters from Élie de Beaumont and by a visit from De Verneuil. I at once saw either that I must defend the position taken up in the opening chapters of the work on Russia . . . or allow geologists to think that the shot you have fired in the eleventh hour was effective in breaking up all my entrenchments. If we had fully co-operated as coadjutors to produce *one work* in Wales and the adjacent parts of England as in the Alps and Scotland, there is every probability that when we found (as we must soon have done) that you had no fossil type essentially differing from the Lower Silurian, the whole of that stage might have been merged in the term Cambrian. . . . I can see no shade of a reason for changing my classification of Upper and Lower. The very limits between them I would select to-morrow if I had to begin afresh. . . . I shall express in public, as strongly as I do in private, my real grief in being compelled, for the first time in our career, to hold any material geological opinion at variance with your own, and shall announce that if you could produce a group of peculiar fossils, I would at once subscribe to your views. . . . Be assured that nothing of late years has annoyed me so seriously as to be compelled to make this defence of the views which I have elaborately worked out, and the classification which I had established after much toil and travel."

The paper which called forth this protest was a continuation of one which Sedgwick had communicated to the Society in March 1845, and of which an abstract was given in the 4th vol. of the *Proceedings*, p. 576. I did not mean to refer, even in a footnote, to any writings on the difference between Sedgwick and Murchison, save those of the two disputants themselves. I am sorry to be compelled to depart from this intention by the reprint of the following passage (from an essay published in 1872) in Dr. Sterry Hunt's recent volume of *Chemical and Geological Essays* (1875). Referring to Sedgwick's paper of March 1845, that writer observes: "That this abstract is made by another than the author is evident from such an expression as—' the author's opinion seems to be grounded on the following facts,' etc. (p. 448), and from the manner in which the terms Lower and Upper Silurian are applied to certain fossiliferous rocks in

Gifted with uncommon vigour of mind and of body, he roused up at intervals to great exertion, and accomplished in a brief space geological feats which would have taken other men a long while painfully to achieve. But these rare powers were retarded by a lethargy, under which he passed long intervals of quiescence,—intervals in which he was probably busy enough with college and other work, but

---

Cumberland. Yet the words of this abstract are quoted with emphasis in 'Siluria' (1st edit. 147), as if they were Sedgwick's own language recognising Murchison's Silurian nomenclature." To this passage the following footnote is added in the reprint : " A letter to the author, written him by the late Professor Sedgwick after reading the above, confirms the opinion here expressed. The abstract in question was furnished by Murchison himself to the Geological Society, the secretary of which declined to receive the abstract offered by Sedgwick of his own paper."

If this last statement was made by the venerable Woodwardian Professor, it may be charitably attributed to the failing memory of age, aggravated by bitter feelings, which were only too likely to distort the recollections of the past. That it is as utterly untrue as it is odious, is proved by the very style of the abstract itself, which is unequivocally Sedgwick's. No two styles could be more readily distinguished than his and Murchison's. The expression which Dr. Hunt quotes *occurs as a footnote*, and was evidently added by the secretary or editor by way of explaining what was not very clearly put in the abstract. The terms "Lower" and "Upper Silurian" are applied in the abstract to fossiliferous formations in Cumberland, which the Professor himself showed by fossil evidence to correspond with what had been called "Lower" and "Upper Silurian" formations in Wales. The expression, "the lower or Protozoic system," occurs at the end of the abstract as a designation for rocks older than and supposed to be marked off by fossil characters from those of the Upper Silurian formations. This use of it was characteristically Sedgwick's (see *Proc. Geol. Soc.*, iv. 223, and *ante*, p. 58), and would have been at once objected to by Murchison, seeing that it struck at the very base of his fundamental doctrine of the unity of his Silurian system. I feel that some apology may be required of me for even alluding to such a charge against a man who from first to last was at least a high-souled, honourable gentleman. Had the imputation remained as it stood in the original essay I should not have deemed it worthy of notice. The subsequent introduction, however, of the authority of Sedgwick, who too was a man of scrupulous honour, seems to demand that Murchison's biographer should indignantly repel a charge which no one would have dared to make while the old lion was alive.

when his real or supposed ailments left him no energy for original geological labour. Perhaps it was partly this pro-crastinating temperament, and partly a naturally generous disposition which kept him so long from taking umbrage at Murchison. In following the events of the next ten or twelve years, however, we must bear in mind that at any moment the peace might have been broken.

From this digression we return to Murchison's work in the summer of 1846. After visits to friends in different parts of England, and among other places to Durham, where he revived his recollections of the Grammar School, the foul drain, the cathedral pinnacles, and the mill-girls or "cotton dollies," as the boys used to call them, he repaired to London to gather together there as many of the foreign savans as had accepted his invitation to attend the Association meeting. Oersted and Forchhammer came from Denmark, Retzius from Sweden, Schönbein from Switzer-land, and Matteucci from Italy.

Not being quite sure whether the Southampton of those days would furnish a large audience to orators in science, he had taken some trouble to give the meeting more popular prestige by securing several leading public men as Vice-Presidents, including the Duke of Wellington, Lord Palmerston, and Mr. Lefevre, the Speaker, besides such prominent names as those of Herschel and Whewell. As a cunning piece of strategy, he induced Sir James Clark to ask Prince Albert to attend, which the Prince very willingly did, not only coming to the President's address, but attending next day at the different Sections. "It diverted me much," Murchison writes, "to see how utterly ignorant the Prince's equerries were of all that they saw and heard, and how

the Prince examined them. I was struck with his mathe-
matical knowledge, for on quitting Section A, at which
Whewell, who presided, gave a demonstration of a new
theorem, his Royal Highness explained what it meant to
his gaping attendants as we drove away to another room.
It was indeed the misfortune of this sensible and gifted man
to be surrounded by persons (as Humboldt foresaw long
before) who were so inferior to himself."

The British Association was now fifteen years old. It
had come through its infancy so well, that there could be
no doubt of its vigorous growth. Nevertheless, some of its
early detractors continued their opposition, to which piquancy
was given by the various ways in which derision and contempt could be expressed. Among these persistent enemies,
the most conspicuous and formidable was the *Times* newspaper, which had followed the Association with the most
uncompromising hostility, refusing at last to print the lucu-
brations of the philosophers unless inserted as advertise-
ments, but continuing its sneering paragraphs or contemptuous
articles. Some of the maligned body felt this keenly. They
could not realize that they had really a ludicrous side ; that
their feasting and holiday-making, their frequent mutual
laudation, and, above all, the opening which their meetings
afforded for any hobby-rider to air his crotchets, were features
which could not but strike the non-scientific outsider, who,
if he could not appreciate the science, might not unnaturally
form but a poor estimate of the usefulness of the Association.
No one of the members winced more under these attacks
than Murchison. Once or twice, indeed, he had written to
an editor either to protest against the spirit of his remarks,
or to correct some error in a statement of fact. Somehow

the Southampton meeting had evoked a renewed outburst of criticism on the part of the *Times*. " Notwithstanding all my efforts and those of my associates," Murchison remarks in his journal, " the meeting was held up to ridicule in the *Times*. But I was nothing cowed, and, at the public dinner at Southampton, I declaimed against such ribald vulgarity and ignorance, saying I was ashamed my eminent foreign friends should go away with the impression that the *Times* in its vituperation of science represented my country, and I vehemently declared that *tempora mutabuntur*. Afterwards, when visiting at Broadlands, I was complaining to Lord Palmerston of the injustice of such treatment. 'Pooh, pooh !' said he, ' never mind them ; a man who is not *Times*-proof cannot succeed in life.' "

Middendorf, the Siberian explorer, who had attended the Association meeting, went with the President to Broadlands, and naturally a good deal of the conversation turned on geographical subjects. " Though Palmerston was Foreign Minister, the house did not abound in maps of distant countries, and when it was desired that Middendorf should give the party some account of his ultra-Siberian travels, no map could be appealed to except an old D'Anville, which Lady Palmerston brought down from a bed-room."

Another house visited at that time was Embly, the seat of Mr. Nightingale, father of the now well-known and universally honoured Florence Nightingale. " Wheatstone was of the party, and he engaged to perform the trick of the invisible girl, by telling you what was in places where no one could see anything. But to do this a confederate was required, and peering into the faces of all the women, he selected Florence as his accomplice, and, having taken her out of the room for half an hour, they came back and performed the trick. On talking to my friend about the talent of the girl, he said, ' Oh ! if I had no other means of living, I could go about to fairs with her and pick up a deal of money.' "

Before returning to London, Murchison extended his visit as far as The Lizard. On the way back he halted at Bath to see his old friend and coadjutor, the late curator at- the Geological Society. " Visited Lonsdale," he says—" Lonsdale's energy in working out a true basis for his coralline descriptions astonishes me, considering his feeble frame, and that he is little more than an *anatomie vivante.* He assured me that it had cost him three weeks to convince himself that the *Stromatopora polymorpha* was a true species. He admits no species which will not bear the test of explaining in its different parts the production of each, the differences between young and old polyps in their cells, thickness of walls, etc. I conferred with Lonsdale as a confidential friend, who was privy to all my progress *in re Siluriana* in reference to friend Sedgwick's tirades. He recommends me to stand firmly to my Lower Silurian, and no one will abandon me, my case being based on plain reason."

With the close of the year 1846 we reach a well-marked point in Murchison's career. He had for some years previously been working with great energy, and at last he had enjoyed the satisfaction of launching the narrative of that work in the *Russia and Ural Mountains.* The success of his labours in science, and the great value of his services to the British Association, had been recognised by his elevation to the President's chair, and he had been honoured by a special mark of favour from his own Sovereign. He seemed

PROFESSOR JOHN PHILLIPS.

From a Photograph by Messrs. Hill and Saunders, Oxford.

to have gained all that in such a career as his was possible, and he might well have been excused had he determined thenceforth to be content with the laurels he had already won. But his activity would not allow him any such retirement.

CHAPTER XIX.

A WINTER IN ROME, AND TWO SUMMERS IN THE ALPS.

IN the early summer of 1847, when the end of the sessions of the Geological and other Societies was approaching, Murchison began seriously to cast about for fresh fields and pastures new. A trip to Cambridge, to visit Whewell, and see the fossil treasures of the Woodwardian museum, then recently arranged by M'Coy, under Sedgwick's supervision, suggested no new outlet for him in British geology. Subsequently, during a short stay in Paris, where among his old scientific friends he met the veteran Von Buch, it was agreed that he should make one of a party to attend the ensuing meeting of the Scienziati Italiani at Venice, and to do some Alpine geology by the way. Out of this proposal there gradually grew a much more extensive plan. Lady Murchison had not been very well since her fever at Tours. For her sake a change of scene seemed eminently desirable, and thus in the end it was resolved to let the house in Belgrave Square, and go abroad for a year or more. In accordance with the geological part of the plan, the eastern Alps would first be visited on the way to Venice. Then a leisurely journey would be made southwards to Rome, there to winter. Next spring might take the travellers to Naples, and lastly, the succeeding summer, if all went well with them, would be spent among the central Alps. Thirty years had passed since that early art tour which Murchison and his wife, in their younger days, had taken through Italy. In repeating it once more they carried with them a new source of enjoyment. In the cities geological museums would now claim, and would no doubt receive, as much attention as the art-galleries used to do ; while in the country, admiration of picturesque scenery or venerable ruin would be mingled with observation of the rocks and their fossils.

The start, however, could not be taken until after the beginning of July, for the President of the British Association at Southampton required to attend the meeting at Oxford, and there resign the chair to his successor in office. If we may judge from his own note-books, this Oxford meeting seems to have brought out, rather disagreeably, one of the points of contrast between the gatherings of the British Association and corresponding meetings abroad. With all the sections sitting simultaneously, and for some four or five hours a day, it is often difficult, or quite impossible, to hear more than one or two papers out of a number which may have been marked out for special attention by the enthusiastic member. And if this member chances to be an office-bearer in- one of the sections, his sense of duty may keep him sitting there, listening, perhaps, to one of the dreary bores who annually inflict their tediousness on the Association, instead of finding his way to one or more of the other sections to hear the announcement of new facts in which he is perhaps specially interested, and regarding which, it may be, he could himself speak a few effective and welcome words. Murchison notes his disgust at these inflictions, and would appear to have found some relief from them in "ordinaries at the Star," "dinners with the dons of Christchurch," "breakfasts in Canon Row,"—in short, his general impression of the best business of the meeting seems to have been pretty much the same as Aguecheek's definition of life, that "it rather consisted of eating and drinking."

After the close of the meeting there still remained another ceremonial before the migration to the Continent could be accomplished. "Early in July," he says, "I went to visit my dear friend George Peacock, Dean of Ely, and met a delightful party, including the Bucklands, Lord Northampton, Willis, etc., and thence we all proceeded to the Cambridge installation, on which day I was admitted by Prince Albert, in presence of the Queen, to be an honorary M.A. S— and I were to have received the honour of LL.D., and our red gowns were ordered, but the caput of the Senate found out that, being neither Lords nor Privy Councillors, we could not have that distinction. S—, who was very proud, was mightily chagrined; but I consoled him by begging him to look at Herschel, Whewell, Sedgwick, and all the clever fellows who, like us, only wore black gowns as Masters of Art.

"At the evening reception by Her Majesty in the Lodge, I presented Struve to her. Whilst walking across the quadrangle of Trinity, Leverrier on one arm, and Struve on the other, we saw the Duke of Wellington coming towards us in his red gown, and I at once said to the two great astronomers, 'Now, gentlemen, here is the opportunity.' Struve was overjoyed, saying it was the thing of all others he wished; but Leverrier turned on his heel and left us, saying, 'Pardonnez, c'est plus fort que moi; je suis Français.'"

After these detentions and a hurried preparation in London, Murchison started for the Continent. His wife had already preceded him, and had been awaiting him for some time at Homburg, where he himself arrived on the 12th of July. As usual, he made full entries in his journal, especially of all objects of geological interest met with in the tour. Nearly a month was taken up in getting to Venice, for the route chosen lay through Bohemia to Vienna, and then in a series of zigzags, to and fro, across the eastern Alps, to see geological sections. The copious details in the journals regarding Alpine geology were partly made use of in the great memoir on the structure of the Alps, to be afterwards referred to. From the other memoranda some gleanings of more general interest will be taken.

Between Murchison and his friend De Verneuil there had grown up a sincere and warm attachment, which, tried under many different conditions, had only grown firmer with each succeeding year of intercourse. They were now again fellow-travellers. Von Keyserling, too, had quitted Russian soil this summer. He joined his former companions at Olmütz, and there the old triumvirate of the Urals met once more. With Barrande and his wonderful collections from the Silurian rocks of Bohemia, they had, of course, much to talk about and to see in the few days spent in his company.

The first entry in the journal after the arrival in Vienna is as follows :—" After seventeen years' absence everything here seems *in statu quo*, except railroads and steam navigation, both, however, great changes for the heavy Austrians.

The hackney-coaches now seem to be more 'langsam' than ever. What a poor capital is Vienna after all! What a mixture of poverty and proud display! What great fat old Swiss porters, and what liveries! What 'Gesellschaftswagen,' like the things our people rode in two hundred years ago. What a mélange of dirty Hungarian peasants and stiff old-fashioned uniforms! What old-fashioned troops, and how slow!—ready to be beaten again by any rapid antagonist."[1]

"*August* 9.—Leave Vienna at 6 A.M. by rail to Vöslau, a small watering-place on the eastern flank of the Austrian Alps, and a little to the west of the high road to Gratz and Trieste. My object was to visit my old antagonist but good friend, Dr. Ami Boué, who after a very ubiquitous and active life, and having done his best to illustrate the geology of Europe from the Highlands of Scotland to Turkey in Europe, is now living a secluded life in this spot."

When the development of geology in Europe is to be traced, the story of the life of this zealous follower of the science will be found to form one of its not least interesting features. Originally of French extraction, his family had settled in Hamburg, where he himself was born in 1794. During his boyhood and youth, when the French armies had overrun Europe, there appeared to be good reason to fear that in the commotion of the times his little patrimony might be swept away. His guardians accordingly resolved that he should have a profession to fall back upon in case of need, and that of medicine was chosen. In pursuance of this plan he came to Edinburgh with good introductions, studied there under Jameson and Hope,

---

[1] This was written before the modern improvements, which have transformed the old city into one of the handsomest capitals in the world.

and took his degree in medicine in the year 1816. By that time, however, the war-clouds had dispersed. It was no longer needful that he should contemplate the possibility of having to live by his profession. He had taken a special liking to geological pursuits, and under Jameson's guidance had acquired considerable knowledge of these subjects. His summer holidays were spent in long rambles over the moors and mountains of Scotland, and in a few years he wrote and published in French his *Essai Géologique sur l'Écosse*, a work full of accurate observations, and in several respects much in advance of its time. Subsequent wanderings took the intrepid geologist over a great part of the Continent, and through some of its most unfrequented paths. He was the earliest pioneer who made known the geological structure of Turkey. His zeal led him to undergo privations of no common kind, and many a time to risk his life. Once he was poisoned and reduced to a state of prostration, from which only a constitution of iron could have restored him. Meanwhile, his pen had been as active as his hammer. Besides several separate volumes, including a valuable guide to the geological traveller, he had written upwards of eighty papers in different scientific journals on many various branches of geological inquiry. We have already seen that he had entered the lists with Sedgwick and Murchison in regard to the structure of the Alps. It was true that he had now retired from active field-work. But his activity in other ways, and especially with the pen, remained unimpaired.

More than twenty years after Murchison's visit, the writer of these lines found the retired traveller at Vöslau, still wearing his old age lightly among the vines which he

had planted over the steep slope of one of the last spurs of
the Alps; keeping yet fresh his interest in the progress of
geology, and well acquainted with its contemporary literature; full of reminiscence of an older time, and showing a
marvellous memory for even the minutest details of places
he had not seen since his early student days; pleased, too,
to use again that knowledge of the English tongue which
he had gained sixty years before, and to find that his work
in Scotland was gratefully remembered by those who had
come after him.   Long letters, written without spectacles in
an almost microscopic handwriting, continue to carry his
friendly sympathy and pleasant gossip to far distant lands,
and serve still as a visible link to bind the present generation with one of the early and almost extinct race of
geological pioneers.   *Floreat in pace!*

From Dr. Boué and his pleasant gardens overlooking
the broad plains of the Leitha, Murchison and De Verneuil
turned southward to strike into the Styrian Alps—that old
geological hunting-ground where Murchison and Sedgwick
had worked so hard.   Railroads had not yet crossed or pierced
the Alpine chain.   Each pass had to be toilfully climbed,
and thus for those who had eyes for them, the rocks, flowers,
and scenery could be looked at leisurely, so as to fix a
lasting impression in the mind.   But now the Semmering,
the Brenner, and Mont Cenis can be traversed easily asleep,
doubtless with much accession of physical comfort, but with
the loss of some emotions, for which such comfort is but a
poor exchange.   In their ascent of the Semmering our two
travellers had time to look at a cutting then being made on
the road, and to note its geological interest, to sketch the
outlines of the dolomite peaks, and cutting off on foot the

zigzags of the road to enjoy the cool fresh morning air of the mountains, yet to reach the summit long before the coachman and his panting horses. In the course of such pedestrian feats in advance of Diligence or hired carriage, much excellent work has been done in Alpine geology.

After so long an interval, and after so intense a devotion to the older rocks, it was pleasant to get back once more to the Secondary and Tertiary strata of the Alps. Murchison led his companion over the Styrian mountains into the Emms Thal, and then across to the deep mountain-girdled lake of Hallstadt, with its village clinging to the cliffs. Of course he could not avoid crossing once again to Gosau, regarding whose rocks he now felt inclined to admit that Sedgwick and he had in some measure been mistaken. And thus revisiting old haunts with more experienced eyes, and passing again through mountain scenery which every succeeding visit can only make more wonderful and impressive, he reached Innspruck, there to rejoin Lady Murchison, and to meet Von Buch, with whom it was agreed to see some more Alpine geology on the way together to Venice.

Leopold von Buch has been already frequently mentioned in the course of this narrative. Murchison had met him often at Berlin, had been some time in his company on the occasion of the Scandinavian meeting described in a former chapter, and had for several years been in frequent correspondence with him. He had formed a very high estimate of Von Buch's powers, of his stubborn energy, and of the almost youthful freshness of his faculties in spite of the passing over him of more than threescore years and ten. He had now occasion during their excursions on foot,

sometimes in rough weather and still rougher wilds, to
learn more personally of him. The notes in Murchison's
journal give us the picture of a man of extreme determination, great perseverance, and a contempt for physical priva-
tion, which furnish a good illustration of the indomitable
spirit which carried Von Buch to the head of geological
science in Germany.[1]

"Von Buch still toddles along (*æt. suæ*, 75) from ten to
twelve leagues on foot. He sent his baggage from Innspruck to Botzen, and came this round in order to see the
Finstermünz Pass." . . .

"Seeing that at Von Buch's slow pace we could not
reach St. Cassian until far in the night, I walked on in a
heavy drenching rain to beat up the Curé of St. Cassian and
get the supper ready. . . . Von Buch has just arrived wet

---

[1] Professor Ramsay furnishes the following note :—" Von Buch was at
the Cambridge British Association Meeting (1845). At Murchison's request
I took him there on the outside of the mail-coach from the head of the
Haymarket. His luggage always consisted only of a small baize bag,
which held a clean shirt and clean silk stockings. He wore knee-breeches
and shoes. Peter Merian, I think, told me that Von Buch in his old age
once started in the afternoon alone from Zermatt to walk over the
Matter Joch (Mont Cervin Pass). They did everything they could to
dissuade him, for there is a glacier-pass to cross, but go he would.
Sometimes people have been lost there in crevasses, and though I crossed
it easily with my wife, yet for an old man to go in the afternoon, and
alone too, was certainly hazardous. So they sent a couple of guides up
after him, who avoided him, and passed him by, keeping away from the
path, out of his sight. Going on a good way they turned back, and walking down the path, met him, and told him that they had been up to cross
the glacier Breuil on the Italian side, that snow had fallen, and that the
glacier was quite impassable. So he turned, and went back to Zermatt,
much against his will. I met him last in 1852 with Merian at Turt-
man in the Rhone valley. He was then quite unaltered, and just as
Murchison describes him. By that time Sir H. de la Beche was very ill,
and unable to walk. Von Buch asked much of him, and said impressively,
' Il faut bien conserver M. De la Beche.' Of all the English geologists
I am sure he respected De la Beche most."

through. He calls at once for his dinner, and will not change his clothes. He eats a hearty dinner with his shoes full of water, and all drying upon him. He was indignant when I begged him to do as I had done, and take the priest's great woollen stockings.

"*Sept.* 3.—We had not gone far on our road to Brunecken, distant twelve leagues, and with a fine panorama of dolomites around us, when it was evident that M. de Buch could never accomplish half that distance. He seemed to De Verneuil and myself to be staggering, and every now and then he sat down on a block of stone. He would, however, hear no reason; said he had often such megrims in the streets of Berlin, and would persist. At a snail's pace, and after a hundred halts, passing through some Alpine villages, he allowed that he was fairly beaten, and at last accepted my proposal to walk on fast to Brunecken, and bring or send back a light calêche to the point to which a practicable road came. So I stepped out once more, passing by the hills on one side of St. Leonhard, with the best of the dolomite peaks on the other.

"Ordering a carriage at St. Lorenzch to be ready for the old hero on his arrival (much, God knows, against his will !) I came on here [Brunecken] to get our Abendessen ready. M. de Buch was suffering from a spasm which had bent him to the right side, and yet he was most unwilling to take any aid from De Verneuil or self, though a little drop of my Kirschwasser seemed to do him good. But what he went through yesterday was enough to damp and cramp the strongest young man. He had previously walked over the mountain passes, and starting at six o'clock on coffee and a crust, and refusing to eat bread and take wine with us at the last Wirthshaus before we came over the Colfosco Alp, he was on foot till six o'clock in the evening, having been from mid-day more or less in the rain, and the last three hours a drenching one, without an umbrella. Arrived at the priest's house, the old man in his dripping clothes (angry with us if we alluded to his state, his hands, which I touched, being icy cold), actually sat for two hours, fortunately in the hot *stube* of the priest, the thermometer outside being nearly at freezing point. He joked, told stories, ate a good dinner, and was up at five o'clock next morning, ready to start in his still moist and damp garments. Hear this, you chamberlains of the Courts of Germany, and imitate, if you can, your brother! This evening, however, he was grateful for the carriage the last two miles. But since his arrival he has quite rallied; has eaten a capital supper of soup, forellen, ham, and pancake, and is now in bed, not, however, before he told us several good jokes.

"If I speak thus of Von Buch, it is only to show his unconquerable spirit and his play of soul. During his morning walk, and when full of pain and spasm, he would every now and then give us a nice little chapter on dolomites and many other things therewith connected."

These "little chapters" sometimes led to lively discussions in which the eager German's impetuosity grew more vigorous from opposition. He ridiculed the modern notion of glacier transport, and halted here and there to plant his staff triumphantly on a big erratic boulder, and energetically demand, " Where is the glacier that could have transported this block and left it sticking here?" Mourning over the spread of such heresies, and looking back with regret to the

creed of the great pioneer of Alpine research, he wrote in
the book of the little inn on Mount St. Gothard:—

> " O Sancte De Saussure
> Ora pro nobis ! "

Murchison thought himself a true and thorough disciple of
the school that preached the doctrine of convulsion and
cataclysm as the origin of the present irregularities of the
earth's surface.   Even now amidst what are everywhere at
the present day admitted to be relics of vanished glaciers,
and with the far gleam of the existing glaciers within sight,
he stuck sturdily to the creed which even the Scandinavian
evidence had not shaken.   " I continue in my old belief,
because I see nothing in the valleys which can be legiti-
mately assigned to glacier action."   In short, in the vast
masses of moraine-rubbish and scattered boulders, he recognised traces only of the powerful torrents set in motion by
such convulsions as those whereby he supposed the chain
of the Alps to have been upheaved.   But in Von Buch he
found a far more thorough-going disciple of convulsion than
himself,—one whose views were too strong even for him !
He writes :—" When Von Buch says that the granite blocks
on the tops of the Jura were shot across the valley of
Geneva like cannon-balls by the great power of the explo-
sive forces of elevation, I feel the impossibility of adhering
to him !"   There must have been a smile on De Verneuil's
face over this schism in the camp of the anti-glacier
leaders.[1]

[1] Von Buch explained the rounded mammilated forms of rock-surface
(now unhesitatingly regarded as *roches moutonnées* due to ice abrasion) by
reference to the concentric exfoliations which many crystalline rocks
exhibit, and whereby a rounded contour not unlike that of *roches moutonnées* is undoubtedly given to exposed knobs or detached blocks.

Among the details of the geological work, there occur in the journal scattered references to the ways and manners of the people. The characteristic devotion of the Tyrolese to their Church, their simplicity and contentedness, their activity and energy, find now and then a passing eulogium. Thus the question is asked :—" Is not the Catholicism of the Tyrol the very best of religions for a good, virtuous, and poor people ? They are full of firm belief, void of gross superstition, with stout bodies, strong heads, and warm hearts. May they ever remain so ! May no cursed economists, may no Miss Martineaus and cotton-spinners ever enter these blessed valleys ! Were I Emperor of Austria, I would forbid the use of the loom in all this glorious country; glorious for its deeds of chivalry and arms, glorious for its thoroughly honest and pious people ! I say it in truth, and before my God, that I would give all I possess to have the faith, and belief, and happiness of these poor Tyrolese."

" Von Buch's broad religious creed is pretty much my own. He states that it was seriously proposed to the Academy of Berlin to discuss the question if governments had a right to domineer the people ? At the same time he admits that the masses must have a comfort and a belief, and agrees with me in admiring the state of the Tyrolese."

While the pedestrian excursions, with their accompanying discussions, were in progress, Mrs. Murchison had been quietly waiting at Botzen until the travellers should rejoin her there. On the 10th September the reunited party reached Venice, which they found full of preparation for the scientific assembly in the following week. It was a new sensation to meet old geological friends among the darting gondolas of the Grand Canal, and to get the latest geological gossip beneath the arcades of Saint Mark. Venice was very full. The old city had determined to give its visitors a good reception, and in its own characteristic way to mingle festivity, fireworks, and boat-racing with the more prosaic business of the men of science. That business, moreover, was to be spread over a fortnight, thus giving a good opportunity of noting how, in Italian fashion, a congress of savans was held. Owing to the arrangement of having different meeting-hours for the sections, a listener could hear all that took place in more of the rooms than one. Availing himself of this advantage, Murchison, after attending the small but select reunion of geologists, and taking part in its proceedings, could mingle with the geographers, or listen to the papers on agriculture and mechanics, or put his head into the doorway of the doctors and apothecaries, whom he again found to muster strong as they had done at Christiania.

One amusing incident of the meeting arose out of a discussion among the geographers and antiquarians regarding the ancient large round map of the world by Fra Mauro, which hangs in the Palace of the Doges. The question as to the exact extent of the Frate's geographical knowledge had branched out into arguments on the old Phœnician voyages, the early discoveries of the Scandinavians, and kindred subjects. One point of doubt and debate centred in " a vessel supposed to be painted on the sea near the rim of this vast round map, and therefore to the south of the Cape of Good Hope, which would prove that the Phœnicians, or some other people, had really rounded the Cape long before the time of Vasco da Gama, whose voyage was forty years subsequent to the map of Fra Mauro. . . . After all this eloquence, a doubt was suggested if there really was a ship painted at the end of the map, and chairs were brought to enable the learned geographers to jump up and decide. It appeared more than doubtful if it was a ship, and thus the section closed! Now what a day we had had of ingenious nothings and misplaced learned verbiage, thought I, as we walked away."

In the same section, on another occasion, "I preached on various heads, chiefly Australian. Citing Siberia and the new countries, and the researches of Leichart, Mitchell, and Sturt, I showed how the north and south chain of Australia would probably prove to be auriferous, because of the same composition as the Ural, etc."

At the close of the meeting a geological pic-nic was organized to explore the Venetian Alps. The party, eleven in number, included De Verneuil, Murchison, "Feld-Marschall Von Buch," as he was sometimes playfully called, Pareto, De Zigno, and other Italians,—a very merry party, which, for about a fortnight, hammered the rocks, told stories, tried to the utmost the resources of unfrequented inns, and finally separated in excellent humour, leaving Murchison to rejoin his wife at Padua, and thence to travel slowly southwards to Rome, where he had arranged to winter.

Once more, then, after the lapse of thirty years, he found himself on the high roads of Italy,—Bologna, Florence, Pisa, Siena, Viterbo, Rome. With very different eyes did he now traverse these well-remembered routes. The sight of the scenes which had kindled his early enthusiasm for art and classical antiquities, renewed again somewhat of the old fervour. But there came with it now a pervading geological flavour, which reveals itself sometimes with an odd bluntness and apparent incongruity among his memoranda; for, while still with an eye for picturesque outlines and interesting sites and ruins, he cannot let any cutting or cliff of rock escape notice—blue marls, macigno, volcanic tuffa, shell-limestone, and other phrases, together with sections of hills and valleys on the line of march, come in among art criticisms or descriptions of landscapes and antique remains. From day to day he enters with laborious detail the geological observations he has made, as if he meant to write an exhaustive memoir on the rocks of Italy., The veriest scrap of a section interests him, and receives at least a passing notice in his memoranda.

This Italian journey had not been long in progress, however, before a new kind of interest arose to claim a large share in the journal. Mutterings of that political hurricane which swept over Europe in the following year, already began to make themselves audible. Italy, from end to end, felt the rising of the storm. Its peoples, separated into artificial nationalities, and, agitated by various and conflicting emotions, furnished a study of absorbing interest to any attentive student of human nature who chanced to live and mingle among them. Murchison, though no politician, could not be in the midst of the excitement without catching some of it and transferring it to his daily record of events. For a while, indeed, his narrative of the sayings and doings of the various parties in the political arena completely usurps the place of rocks, fossils, art, and antiquities. Some extracts will serve to show in what a stirring time his second Italian sojourn was passed, and how shrewdly he could sometimes note events as they occurred, or as they loomed dimly in the future.

It was at Florence, as early as the middle of October in this year, that the first indications of the coming troubles

met him, and he thus prognosticates :—" People who know
the Florentines tell me that all this *vox populi* will subside
with the first *tramontano*, or cooling north-west wind, but I
put little faith in their sayings. An enormous change has
been made in the whole framework of Italian society, of the
ultimate effects of which I am by no means certain. I
cannot see the mighty assemblages of singers in the streets,
nor their enthusiasm, the crowds of drunken people in the
low *cafés*, and the complete license which they are taking,
without a fear that the revolutionists, that is, root and branch
fellows, will work something of their own out of it, and will
shake, if they do not eventually upset, the throne of their
benevolent sovereign, the Grand Duke."

Things were beginning to look very dark at Rome. One
of Murchison's first remarks on his arrival there is as follows :—" Rome is fallen away from what I left it thirty
years ago, and a single walk round my old haunts led me to
see that in beggary, filth, and decay, she is more pre-eminent
than ever. What a singular fatuity in those who govern to
expect to produce effectual reforms, when all is as rotten at
the core as the mouldering antiques, whose foundations are
daily giving away under the old edifices ! A truly efficient
reform would be to retrench one-half of the overgrown charges
for cardinals and priests, employ the poor, and reanimate
this land of the dead."

Again he breaks forth :—" Oh for an Oliver Cromwell to
drive this ermined vermin from the world, or, at all events,
to subordinate them to a good civil government ! Yet here,
forsooth, it is that we hear of the *risorgimento dell' Italia*,
and such nonsense, amidst squalor and rags, and with a
hundred beggars at the side of every carriage. When is the

day of retribution to come? Not, I apprehend, for many a day. There is no intelligent middle class. I expect, however, scenes of great disorder and tumult, followed by a good deal of highway robbery, and the revival of the good old times of the brigands.

"How is improvement to be combined with the conservation of the Papacy? How is the spiritual power of the Pope to be untouched? How is His Holiness to be left in undisturbed possession of the influence he wields over all Catholics, foreign and domestic, whilst in his own States, the laws, internal government, trade, commerce, etc., are to be administered by civilians? That this separation will be attempted sooner or later no one can doubt, now that His Holiness has put arms into the hands of many thousand citizens, two-thirds of whom will seek for such an adjustment, and will eventually compel it."

But, apart from this temporary political aberration, how thoroughly geology had now engrossed Murchison's mind is well illustrated by what he says of himself a day or two after he had settled down in Rome. "A visit to the Vatican revived some of my foregone pleasures; and glorious bright gleams over all the Campagna and the distant snowy mountains, with the sea glittering at Ostia. These and the finest sunsets from the Pincian are not enough for the unhorsed geological knight. For here, in truth, I find myself a fish out of water, an animal without belongings, and deprived of the conditions in which I have lived for some years past, viz., a set of men with pursuits entirely akin to my own. The other avocations of the sight-seer in the Eternal City are forced on me, *faute de mieux*, and I endeavour with these, and visits to the studios of the Tene-

rani and Gibson, to occupy the mind, but in vain. These
are only *scherzi.* My geological note-books of the summer
lie undigested, and I lack the courage and stimulus in this
city of indolence to work them up into something. In
short, that something on the Alps and Apennines would
be too general and desultory, and too little decisive to
satisfy me. I feel this, and must desist at present from the
endeavour. I wished to be in Sicily and doing something,
but the troubles stopped me ; and now I find I might have
gone there and explored the mountains before the snow fell.
I look daily to the Apennines with a wish to be there,
but their backbone is a mass of snow, and I should only do
things by halves until I can make clear transverse sections
of the whole chain. I may, however, make a run to Tivoli
and the Sabine Farm of Horace, and compare the rocks with
the version given of them by Ponzi. But this is fragmentary
work, and unworthy of me. Often do I wish that the
long quarantine of winter were passed, and that I was once
more at work with my hammer, and out of the gulf-stream,
of English sight-seers and ' Syntaxes ' in search of the picturesque. And yet I have joined an English club here, where
Lord L—, gouty and unable to stand on his legs, talks of
his shooting pigeons and larks from his carriage, whilst a
lot of young dandies are betting on their horses and their
performances in red coats. As a. climax, I am just going out
with Canino [Charles Buonaparte] mounted on one of his
nags to view the throw-off at Magnonella on the road to
Civita Vecchia."

After this tedious fashion, with an occasional country
excursion and a peep at the rocks, such as they are, of the
Campagna, Murchison contrived to get through the winter

at Rome. With the earliest advent of spring he started for Naples to renew his acquaintance with the volcanic aspect of that region. But while the season of the year brightened from mid-winter towards summer, the political horizon grew each week more ominous. Disaffection reigned everywhere, breaking out abundantly into open revolt; governments in terror unable to act, troops fraternizing with insurgents, crowned heads seeking safety in flight, communism and democracy rising on all sides and threatening to supplant every constituted authority. From Vienna, Berlin, Paris, each post brought more exciting news, which added fresh fuel to the fire of Italian discontent. Naples was no longer a very safe or pleasant place for a stranger to stay in. Murchison remained long enough to witness the expulsion of the Jesuits in obedience to popular clamour. He thereupon, after getting as far south as Sorrento, turns northward again. On the 12th March we find him writing thus :—" If I croaked about the prospects of the Italians in their new condition, and foresaw in the establishment of civic guards the germ of the overthrow of all monarchical government, how forcibly is this conviction driven home now that the French Republic is re-established by the zealous co-operation of the National Guards of Paris ! This is what I have foreseen for years. I have always said that in the formation of the National Guards, France had established a body of janissaries who would change dynasty and government at their will and pleasure." . . .

" If the detestable law of the division of property, which deprives an industrious individual of the power of establishing a family, and forbids the preservation of an old lineage were to be applied to England, then should we proud islanders

become in a trice Yankees, and with the addition of National
Guards our ruin would be complete. I now begin to think
that Tocqueville was right in his view of the rapid spread of
democracy."

On his return to Rome the geologist found matters so
serious that, in the curiosity to see and hear everything going
on, he forgets for a little his geological exile.

"*March* 22*d.*—The news about Vienna set all the Romans
agog—rising of the Viennese, expulsion of Metternich, formation of a new government, capture of the arsenal after much
loss of blood, imprisonment of the Imperial family, *cum multis
aliis*. Pistol-shots were soon heard in all directions, bells
were ringing, and every one moving about crying, ' Festa,
festa !' In less than two hours all the Corso was hung with
banners, and about one or two o'clock a crowd of no great
numbers (some say not more than fifty) applied ladders to
the Palazzo di Venezia, and dowsed the arms of the Austrian
ambassador. These worthies were, it is said, chiefly 'stranieri,'
and a Lombard mason was he who chiefly distinguished himself. The civic guard, one of whose posts is opposite, looked
on, and no sort of remonstrance was made by any authority
of police or otherwise. The same parties took down the
arms from any other palaces or houses on which the Austrian
eagles were quartered. I saw a set coming down, not above
twenty persons performing ; others, as at the Ruspoli, took
them down of their own accord. When the rioters entered
the Palazzo di Venezia, the gates of which were quite open,
the old ambassador, Count Lützow, on their walking up-stairs,
demanded their object, and they replied that the Austrian
Government being dissolved, he ought to take down his arms,
just as the Frenchman had done. Against this Count Lützow

protested, and they then went to work, the Lombard mason operating in what Masi, the poet of Charles Bonaparte, styled a truly dramatic manner !

"I was driving up the Corso with my wife to see the hangings and draperies of the houses, when we were stopped by a crowd, carrying with them the arms of Austria, and so it was 'volti subito;' and whilst they were thus transported with military music (a regular band of the Civica attending) we drove to the Pincian, and there overlooked the bonfire which was made of the imperial ensigns, around which a guard of the Civica played as the flames rose up; young and old *gamins* jumped upon the planks and paintings when first thrown on, and thus the orgy went on, about three or four hundred people only being armed, and not fifty regularly operating. A platoon of London police would have set them all to flight, and a company of English guards would have restored order to all the city. But officers and soldiers of the Civica all looked on, and all was 'allegrezza.'

"This concluded, I went on foot to the other end of the Corso, opposite the Palazzo di Venezia, and witnessed the Roman demonstration of several thousand persons, who came by arm-in-arm, including a great number of hungry-looking wretches, mixed with citizens with good coats, young Italians of all grades, civic guards, soldiers of the line, both horse and foot, and even a few ecclesiastics and women. They had tricoloured banners of all sorts and sizes, and they screeched still louder with their songs and hymns as they passed the ambassador's residence, on which the Pope's arms remained, and in place of the arms of Italy a tricoloured flag, with the words 'Alta Italia!' in great letters. Parading by, but not attacking further, they went round and made some furious gesticulations before the church and houses of the Jesuits. Among the crowd was young O—, the son of the Neapolitan ambassador; so completely has the ' amor Italiæ' maddened every youth.

" Then came the vespers and the Moccoletti, when all the Corso and many parts of the town were illuminated, and the people walked and ran up and down the Corso, each with his little taper. The sight was grand and curious. In the meantime, before dark, one of the leading zealots, whose name I forbear to mention for old acquaintance sake, had a large white placard posted on the east wall of the Palazzo di Venezia, with the words ' Dieta Italiana !'

" The shoutings and vociferations of the squads, with their various banners, during the Moccoletti, and still the order with which all was conducted, were truly striking. Rows of women stood on the raised flags which flank the Corso, with carriages here and there. After about an hour and a half of this nonsense, a cry ran from one end to the other, and in two minutes every light was extinguished, and the people filed off to their houses, charmed with their fête and their revenge.

" Nothing surprises a foreigner so much as the union and celerity with which these festive rows are got up. The mandate of the fiery cross could have been nothing to the prompt stage-like evolutions of young and old ' Romani.' When such a play has to be enacted every one is in his part, all singing in harmony, and all knowing the words."

" *Rome, April* 2.—Yesterday, April the 1st (ominous), was a great jubilee here, with illuminations taking place in the evening, on account of the recovery of the head of St. Andrew, which was stolen from its resting-place in

St. Peter's, to the great horror of the Pope, who, with many of his flock, had viewed this theft of the holy relic as a sad omen. This head was richly bedizened, and the thieves having taken from it all that was valuable to them, buried it outside of the gate of the Cavalli Leggieri, where, being dug up, it was carried in an ecstasy of joy to the Pope. My ' laquais de place' Ramieri says, that the loss of the ornaments is nothing, the great point being the recovery of. the head and the *'cervelle'* of the Saint. As this St. Andrew was said to have been brought from the Eastern Greek Church territories, the common people had suspected the Emperor of Russia (now in bad odour) 'as the thief. I wonder it did not occur to them to lay it at the door of some canny Scot, who sought to sanctify the Land of Cakes in these perilous times ! Would any rational being believe that this is one of the many dramatic scenes in the *risorgimento dell' Italia*, and has followed close on the expulsion of the Jesuits !"

Now and then, in the midst of this unwonted devotion to political events, there occurs an allusion in the journal to the science which, for the time, those events had displaced. Thus :—" Geology, Rome. Well is it to turn from the dark vista of communism and destruction with which the political horizon is shrouded to Nature and Nature's works. But alas ! in Rome and around it there is little to be done. A ride the other day with M. Louis enabled me to review the succession' of the tertiary and volcanic rocks on the right bank of the Tiber, *i.e.*, in all the tract between Ponte Molle and the Porta dei Cavalli Leggieri behind the Vatican."

Nevertheless, in spite of these various impediments, some pleasant excursions were made, partly geological, partly antiquarian, up the valley of the "præceps Anio," round by the hills of Latium, and over the heights of Albano. It was during these rambles that the materials were gathered for the descriptions afterwards given of the geology of that part of Italy to the Geological Society.[1] It had been Murchison's intention to explore the Apennines and cross over to the eastern side of the peninsula, for the purpose of examining the cliffs of the Adriatic; but in the disturbed state of the country such a tour was found to be impracticable. Accordingly, he at last broke up his winter camp in Rome and turned northwards, to attempt that further Alpine campaign on which he had now set his heart.

"*Rome to Narni, May* 1.—Thirty years ago I travelled this road with my same old wife, then a young one. The Campagna is indeed the same, but I see it with very different eyes now,—that is, *geologicè*. Formerly I talked of the rocks of Salvator Rosa, now I know that Salvator Rosa was a gross humbug. Mr. Ruskin is quite right in his dictum that Salvator's rocks are all inventions or scene-paintings for those who never studied nature.

"At the gate of Rome the 'civici' (animals I detest) looked sternly into our carriage, to be certain that my wife was not a cardinal in disguise! For the liberty of the subject being now assured, the Holy College is incarcerated in Rome, to answer for all former sins. The Pope has published his Latin allocution, and has abjured the Roman wars which his subjects and ministers have waged in spite of him, and which he opposed 'quantum in nobis fuit.' . . .

---

[1] On the Structure of the Alps, *Quart. Jour. Geol. Soc.*, v. p. 280-3. On the Earlier Volcanic Rocks of the Papal States and the adjacent parts of Italy.—*Op. cit.* vi. 281.

Breitenberg (in continuation of the Stauffen ... S. of Bregenz)

S

N

Tertiary rocks
sandstone & dirty
against the
Nickel rocks.

Haslach
Bahn

Blacks

Great torrent wall

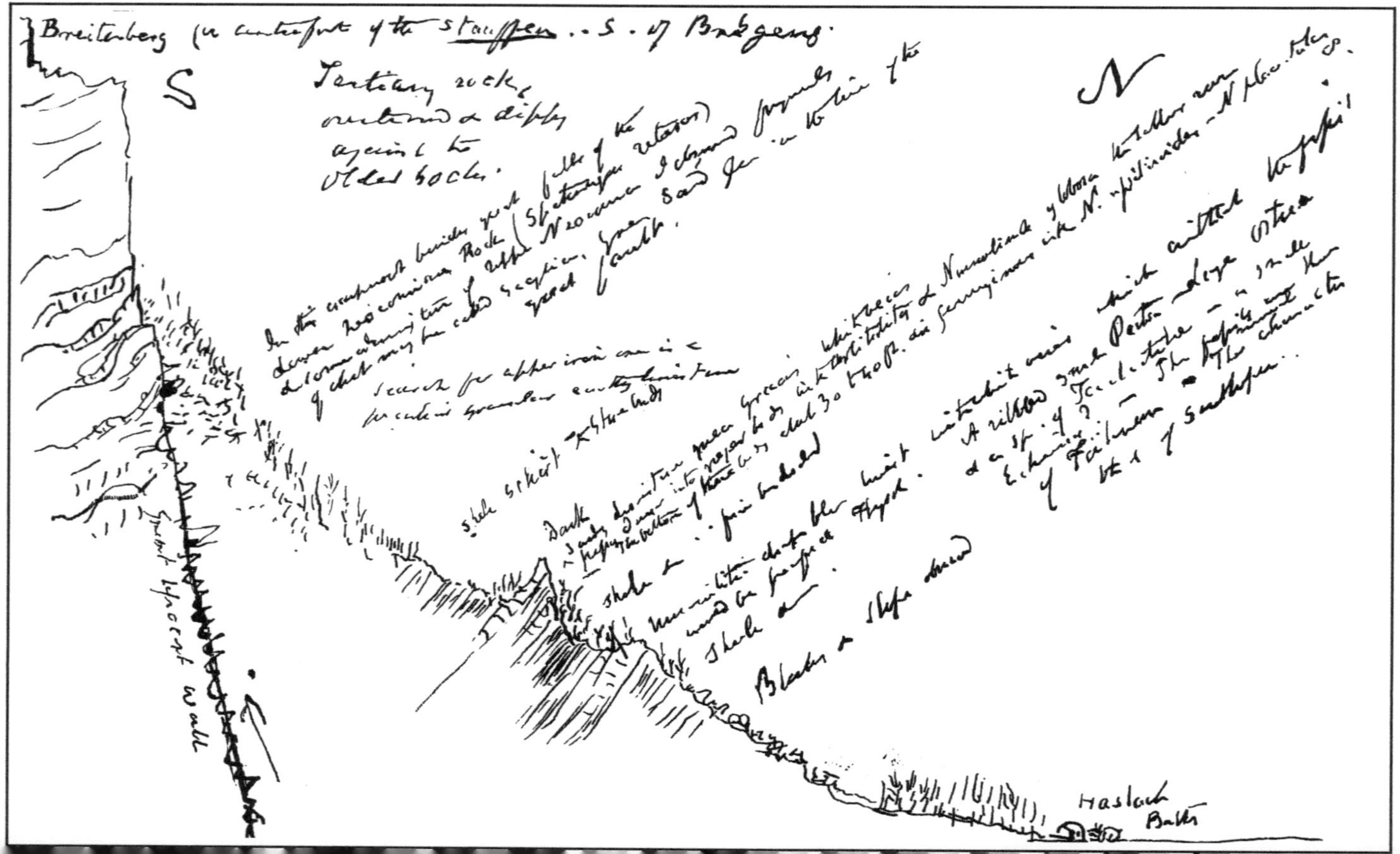

" Abandoning the deplorable politics of Rome, I note that
the tract around La Storta, the first post, is all composed of
flattened and well-arranged subaqueous volcanic tufa.　Bac-
cano, as before, seemed to me anything but a centre of eruption." And so his notes run on, politics, geology, scenery,
and antiquities coming in each for its share of comment.

Nearly five months were consumed in the homeward
journey, most of it being spent among the Alps; not in mere
sight-seeing or lazy admiration of fine scenery, but in hard
bodily toil.　Back and forward, traverse after traverse was
made of the Alpine ridges, valley after valley was explored,
and lines of geological section were drawn at intervals across
a great part of this chain.　Beginning with a tour on horseback through the northern end of the Apennines, Murchison
struck into the main chain of the Alps by the pass of Mont
Cenis, crossing it to Chambéry, thence to Chamouni, and down
to Geneva.　A sojourn was once more made at Vevay, among
Lady Murchison's Swiss connexions, after which the geologist started alone to make another section of the Alpine
chain by way of Val de Ferret and Courmayeur to Aosta, and
back by the Saint Bernard to Martigny and Vevay.　The
Swiss scientific congress was held this year at Soleure, and
Murchison attended it, meeting there the leading Swiss
geologists, with some of whom he turned southwards into
the glacier valleys of the Oberland.　The later stages of the
tour included much climbing and detailed work around the
Lake of the Four Cantons, the Hohe Sentis, Sonthofen, and
Bregenz (see Plate III.), until at last reaching Basle the
travellers sped down the Rhine, and with two or three halts
by the way to see old scientific friends, finally landed in
England.

The journal of this prolonged Alpine tour consists mainly of geological notes, which were afterwards worked up into the memoir on the Alps, already referred to. Two objects seem to have been constantly in the traveller's mind: 1st, To get hold of the structure of the mountains and disentangle the true succession of the crumpled and fractured groups of Alpine rocks; 2d, To see for himself the nature and value of the proofs so confidently adduced by his opponents in the glacier-iceberg controversy as evidence of the former vast extension of the present glacier system of the Alps.

So thoroughly had he wedded himself to the iceberg hypothesis, and so prejudiced had he grown against any effective geological work on the part of glaciers, that even now, in the very stronghold of the glaciers, he refuses to admit them into the list of powerful terrestrial agents. Floating ice and débâcles caused by the convulsive upheavals of the mountains will, in his opinion, account for all the phenomena. When he meets with vast piles of detritus in some of the valleys, now generally recognised as re-assorted moraine stuff, he notes that it "would doubtless be called moraine by the ice-mad folks." The advent of such a disbeliever into the very sanctuary of the glaciers provoked discussion, for he rather boasted of his opposition. At Aosta, for instance, the worthy meteorological Chanoine Carrel, to whom glaciers and their doings had long been familiar, answered the adverse argument with the easy nonchalance of a man who felt that argument of any sort was needless. "The Canon," says the journal, "fires his shot as dreamily as if all were true as gospel. 'We have striated rocks, and polished surfaces and blocks. Nothing but ice could ever polish, striate, or transport. Hence all was ice down to the lower valleys.' If you say to him, 'Then all the plains of Northern Europe were ice-clad also; and if so, all Europe;' he replies coolly, 'And why not?'" This was too much for Murchison. But it was nevertheless true, and he lived in some measure to acknowledge it.

With another and more noted champion of the glaciers De Charpentier, our traveller entered into further discussion. That shrewd observer had clearly shown the spread of the huge erratic blocks over the plains of Switzerland and along the slopes of the Jura, and had connected these blocks with a former wide extension of the glaciers. Murchison visited him at the pretty hamlet of Devens, and went with him to see the famous blocks of Monthey and other travelled boulders of the neighbourhood. Charpentier, in explanation of his views, drew a little diagram in his companion's note-book, showing how the highest erratics had been stranded by the greatest spread of the ice, and how, as that ice shrank, the stones were lodged at lower and lower levels. Murchison adds a note to the drawing: "This little diagram, drawn by the hand of M. Charpentier, shows his ingenuity, but does not convince me."

His scepticism could hardly be accounted for by want of acquaintance with glaciers. In this very tour he spent much time among the ice in the Mont Blanc group of mountains, making some good ascents with Auguste Balmat, and keeping his eye ever open to the reception of evidence of the work which land-ice actually performs.

The gathering of naturalists at Soleure is thus chronicled:—" *July* 24.—Yesterday evening, being Sunday, the savans came in quietly, without beat of drum, as the country girls and farmers, who had been into town in their best, drove

out of it, and we had a supper under our rooms in the Crown.
The geologists here are Studer, Lardy, Hugi, Montmollin,
Favre, Dubois de Montpéreux, Escher von der Linth, and
P. Merian. Here I made the acquaintance of young Rüti-
meyer. . . .

"A capital finish to three very instructive and agreeable
days at Soleure, under the presidency of the kind and highly
respected apothecary, Pflügel, a septuagenarian. The old
man has scarcely slept sound for the last year in expectation
of the honour, and in fear of not living to accomplish his
duties. He gave us a *fête champêtre*, in illuminated tents,
at his little villa overlooking the Aar. Then the students
marched up to our table singing their national hymns, and
each carrying a coloured light, whilst the good old man circulated ever around us to do the honours. It was a scene
never to be forgotten. At our first public dinner I had to
return thanks for my health, proposed by Schönbein with his
usual animation and originality. At our last feast I proposed, after a bit of a speech in French, with allusion to
Agassiz and other Swiss, 'Perpetual prosperity to the Swiss
Society!' which was drunk with a loud 'Hohe' and raptur-
ous applause, all the savans bringing their glasses up to me
to tingle-jingle. I requested them to sing the air of the
'Vaterland' to the tune of our 'God save the Queen,' and
we had a jolly chorus. Peter Merian, of Basle, then made
a humorous speech worthy of Peter Robertson."[1] These
jocular proceedings continued after the meeting, for the
choicest of the geological spirits went with Murchison up
to Grindelwald, and formed, as he remarks, "a very merry
party."

[1] A well-known Scotch lawyer and wit.

On the way down the Rhine valley, and during the short halt at Bonn, chance once more brought Murchison and Von Buch together. The veteran geologist, none the worse of his recent Alpine sufferings, was busy with the study of the Chalk and the Nummulitic Limestones, throwing into it all his youthful ardour. He listened to Murchison's narrative, and especially to that part of it relating to erratic blocks and glaciers, and no doubt ministered some comfort to the narrator's mind by standing out, as of old, for torrents of water sweeping everything before them at the upheaval of the Alps.

On the 23d September, after an absence of about fourteen months, the travellers resumed their place in Belgrave Square.

CHAPTER XX.

THE COPLEY MEDAL.

IN less than two months after his return to England from his prolonged tour of 1848, Murchison produced at the Geological Society one of the longest, and what was considered by some of his friends to be perhaps the best of all his original memoirs.[1] In this elaborate paper he combined his recent observations among the Alps and Apennines with those of a former year among the Carpathians, his special object being to prove a transition from Secondary into Tertiary rocks, and to show over what a wide extent of Southern Europe, and in what massive proportions, rocks of Eocene age extend. In early days Sedgwick and he had done good service in showing how the vast Secondary deposits of the Alps range upwards into older Tertiary masses. They had erred, indeed, in some respects. Their contention as to the age of the Gosau beds had not been sustained by an appeal to facts. Nevertheless, Murchison had felt sure that they could not have been so wholly mistaken as Con-

[1] On the Geological Structure of the Alps, Apennines, and Carpathians, *Quart. Journ. Geol. Soc.*, v. (1849), 157-312, with plate of sections.

tinental geologists had supposed, and that this conviction of
his had been well founded he now proceeded to demonstrate
in this paper. By the evidence of good physical sections
and of fossils, he showed that instead of a meagre development of the older Tertiary strata, there existed in the chain
of the Alps, and throughout the South of Europe, enormous
masses belonging to that geological series. The hard, greenish
sandstones and schists, which form such notable mountain-
ridges along the flanks of the Alps, he showed to be of
the same general geological age as soft clays and sands in
the north-west of France and the south-east of England.
Undoubtedly he availed himself to the full of the assistance
he had derived from the geological friends with whom he
had examined the ground, and without whose local knowledge indeed he would have been comparatively helpless.
But he as fully acknowledges the obligation. It was no
small merit to bring all the scattered observations of many
different students along the great line of the Alps into
relation with each other, and to make them lend their aid
in bringing out some essential features in the architecture
of these mountains.

The judgment formed of this essay in Alpine geology
by a very competent judge is flatteringly, but doubtless
truthfully, expressed in the following extract. Von Buch
had hitherto been in the habit of writing to his friend in
French. He now begins a letter in English, but after the
first paragraph (here reproduced), relapses into the former
language. His English, however, as the reader may judge,
was quite as good as his French, and, indeed, showed no
common mastery of a foreign language, expressing vigorous

thought in vigorous words:[1]—" *Berlin, 20th December* 1850.
—MY DEAR SIR,—Your admirable paper on the Alps has
always been my companion during my rambles last summer
in Switzerland.   Every day when I took forth my breviary
I could not help to repeat ' Je vous admire.'   Such a genius
of sound and extensive combination, the very test of an
eminent geologist, was never before ; such talent of exposition will always be a very rare and admirable gift.   You
do approach the Nature to lift up her veil with due reverence and attention to her, and then she speaks to you
graciously.   Others come hastily with spurs and boots and
gross hands to draw the veil, as it was a curtain, and they
discover behind not the flying nature, but a phantom they
have constructed themselves.   Such are the makers of coral-
islands[2] dancing up and down on the sea, the builders of
volcanic cones by successive lava threads,[3] and so many
other ingenious ' explainers ' of nature.   Your ' Alps ' will
rest for all times a model of investigation. . . . —Believe
me, my dear Sir, your faithful admirer,

                                            " LEOPOLD VON BUCH."

[1] No apology seems needed for giving Von Buch's letters *verbatim.*
Some of their characteristic features would be lost by any correction.
And even with their slips they are very remarkable productions to have
been written by so venerable a foreigner, who acquired his knowledge of
English at a time when the language was not so much known on the
Continent as it has since become.

[2] Darwin's admirably worked-out theory of the formation of coral-
islands, now so generally received as a sound and firmly established contribution to scientific reasoning, evidently found no favour in the eyes of
this dear old geological Tory.

[3] He refers to the explanation which accounted for the growth of
volcanic cones by the accumulation of successive eruptions of solid
materials round the focus of emission.   His own famous doctrine was
that of the " Erhebungs-Kratere"—that is, that volcanic cones are so
many huge blisters raised on the surface of the earth by the outward
swelling of the molten masses within.   Though sturdily maintained by

Whether the rapid preparation of his great memoir, or the transition from Continental life to the more exacting calls of London had told on his health, may be doubtful, but now, for the first time in this biography, we meet with Murchison as an invalid, and even somewhat of a valetudinarian. Lady Murchison, too, was ailing. So in the summer of 1849 they spent some time at Buxton and other watering-places, and in visiting friends, with the view of getting rest and renewed health. The geologist took very unkindly to this change in his occupation. He had never been used to think of his bodily frame except as a machine for carrying his indomitable spirit from place to place. But now his note-books and letters assume a hue not unlike that of his friend Sedgwick under the most depressing hypochondria. "How dull, tame, and insufferable," he exclaims at Buxton, " is the west midland geology after Alpine frolics ! and what a gloomy and sepulchral air has every English watering-place after the baths of the Continent !" The gossip of the baths and the reminiscences of old Peninsular comrades formed but a sorry exchange for the scrambles with De Verneuil, Von Buch, or Escher von der Linth, up the slopes of the Hohe Sentis or Tyrolean Dolomites.

He had planned several pleasant rambles during the summer. For instance, he wrote to Phillips :—

" 28th July 1849.—This watershed of England is perhaps the best place during the cholera-plagues, and I wanted good air and quiet. I hope to be soon as sound as a geologist ought to be in the summer, and then think of excursionizing a little before the Birmingham meeting, and I wish to consult

him and his followers, and espoused by some geologists of note, especially by Élie de Beaumont, it has gradually died out, and the opposite view, which he derides in the above letter, has prevailed.

you about this trip. My present notion is to join Ramsay, if he is, as I suppose, somewhere in North Wales, there to learn, as well as I can, the progress that has been made since the bygone days of Silurianism. I really wish to bring myself up to the existing state of knowledge in these beloved regions, and a few mountain-walks will, I trust, complete my cure. *Inter alia*, I shall take a glance certainly at your Malvernia, of which you have rendered every corner so attractive."

But a few weeks afterwards his tone alters :—

"*August* 19, 1849.—MY DEAR RAMSAY,—I never wrote a letter with more repugnance than when I inform you that I cannot join you at the foot of Snowdon. In truth I am (I regret to say) quite unfitted for any geological excursion on account of health and nerves. I have been more ailing this year than in any previous one, and I have not regained what I expected at Buxton.

" This is the first summer in my life in which I have been fairly obliged to strike work, and the very thought of it adds to my depression. Hoping, however, that I may be in better plight when the Birmingham meeting comes off, I trust to meeting you there, and who knows that later in the autumn I may not accomplish what I have failed to do now under your kind and instructive auspices."

By the time of the assembling of the British Association at Birmingham in September, he had so far recovered himself as to be able to attend the meeting. The proximity of the Silurian territory gave the geologists an opportunity of making an excursion into it under Murchison's guidance. A goodly gathering mustered round him as he led them into the Dudley caverns, which Lord Ward lighted up for the occasion. There he gave them a subterranean lecture on the story of the ancient sea-formed rock within which they were assembled, and then emerging again to daylight, he led them to the top of the Wren's Nest, where, amid much merriment, and with general approbation, he was enthroned by the Bishop of Oxford as " King of Siluria." [1]

But immediately after the meeting we find him at Brighton drinking mineral waters, and noting each disor-

The South End of the Wren's Nest, Dudley.—(From a sketch by Dr. Whewell, *Sil. Syst.*, p. 485.)

dered symptom about him, with the treatment recommended, as if he had been all his life a confirmed valetudinarian. A man whose chief subjects of thought at the time would seem to have been the miseries of " suppressed gout," " ill-arranged bile," " stomach attacks," "vertigo," and "bad nights," with the relative effects of " bismuth pills," " blue pill," and " cordial rhubarb ;" and who, after trying in vain the virtues of " Carls-

---

[1] These doings at Dudley formed the subject of a string of doggrel verses by some local bard, published at the time, with the title of " The Dudley Gathering, a Ballad." This was the second time that Murchison had given an underground lecture in the Dudley caverns. At the former meeting of the British Association in Birmingham (1839) he had led a similar party of excursionists. See his description of that earlier visit in a letter to Hugh Miller, in Bayne's *Life of Hugh Miller*.

bad waters," " resumed tippling the waters at the Ems tap "—
such a man could hardly be expected to produce any coherent mental work. A day's shooting now and then was tried,
particularly in that old haunt, the pheasant-covers of Up
Park, but with no good effects. " What a weathercock," he
writes, " is a bothered stomach, affected as mine is, through
every pore of the skin, and how unnerved is the stout man
of yesterday! It is freedom from these ailments which is the
basis of the success of great public men—an iron stomach,
the skin of a pachyderm, and no nerves, forming the *sine quâ
non* of a Duke of Wellington."

In the brighter intervals of this season of despondency,
plans were laid for further scientific work. Thus : " Began
to prepare a memoir on the pseudo-volcanic rocks of Italy,
more to revise the subject and keep my hand in than to
give out many new observations. But I must speak my
mind on several points of Tuscan, Roman and Neapolitan
igneous action, and on the succession of the events. Also
began to speculate on a general Geological Map of Europe.
With this sort of serious occupation, much idling, and a
good deal of open air promenading—the weather turning into
a fine Martinmas summer—I have eked out my time."

While still in this unwonted state of nervous depression,
he received the announcement that the Royal Society had
unanimously assigned to him the Copley gold medal. The
terms of the award ran thus :—" That the Copley medal be
awarded to Sir Roderick Impey Murchison, F.R.S., for the
eminent services he has rendered to geological science during
many years of active observation in several parts of Europe,
and especially for the establishment of that classification of
the older palæozoic deposits, designated the Silurian system,

as set forth in the two works entitled *The Silurian System founded on Geological Researches in England,* and *The Geology of Russia and the Ural Mountains."*

By none were his claims to the honour more urgently pressed than by his generous old friend Sedgwick, to whom he thus wrote at the time :—

" *Nov.* 23, 1849.—MY DEAR SEDGWICK,—I have lighted my candles to write to you in an orange fog, and to thank you for your letter of yesterday, and still more for your warm espousal of my claims to the Copley medal, of which I heard from Horner when he wrote to announce the result. That result, prefaced as it was by the cordial support of all the geologists, is unquestionably the highest honour I have ever received, and the more gratifying from the manner in which the vote passed unanimously through the R. S. Council. Airey, in writing to me about it (and he was the proposer), said it was carried *nem. con.* and *nem. tac.*

" My wife and self are both very sorry to hear of your poor health. We are all in the same boat. For the first two days after her arrival here, she has been a sufferer. And at times I say of myself that I am but an old woman. But the Copley is a sort of cordial-ball for an old poster, and I must revive, if possible. I knew very well the causes of your silence without your repeating them. There are but few friends on whom I could count to the death, and that you, my dear Sedgwick, are one of them, is one of the things I am most proud of."

Gradually regaining his former health and spirits, Murchison resumed by degrees his place in the midst of the activity of the learned Societies in London. He had been appointed a Vice-President of the Royal Society, and was oftener at its meetings than before. The following extracts from his correspondence show how thoroughly he had at last got once more into harness. To Sedgwick he writes on 18th January :—" If you rally quite, and ever intend to be at our Society this winter, I should like to bespeak you for the 6th February, when I give forth on the pseudo-volcanic rocks of central Italy, and when we shall necessarily go more into the question of elevation craters which Lyell has reproduced. I am convinced that the truth lies neither in the extreme of Von Buch and De Beaumont, nor in that of Lyell. . . . I suppose we shall give our Wollaston medal this year to Hopkins, and it could not go to a better man.

" I was in the chair of the Royal Society last night as one of Lord Rosse's V.-P.'s. . . . The paper on the alkaline bases, by Dr. Hofmann, astonished me, and proved to me (an old pupil of Brande and Faraday) that I was incapable of understanding the elements and phraseology of the science as it is now carried on. As a mathematician, you could do so with a little labour ; but the *isomerism* of ammonia, in which all the ammonia escapes, puzzles me sorely. It is another world of science."

To Phillips, on 9th February :—" I suppose you know that at the last anniversary of the Royal Society it was announced that Lord John Russell would give £1000 per annum for the advancement of science, to be at the command of the Council of the R. S. *Entre nous*, I wish that the Premier had divided this thousand into two parts, and had given one to our great national patriarch of science, the British Association, and another to the corporate metropolitan body. But the offer has been made spontaneously on his part, and it was not for men of science to repudiate it. They have put me on the committee of the R. S. to prepare a report on the objects proposed to be carried out by the grant. . . . It occurs to me that we must fly at game at least as noble as we of the B. A. have attempted to take. In truth, the minister will, if he carries out this project, to a great extent disarm the B. A., which, if it have no funds to dispose of, will simply become a pleasant rendezvous for a few men of science on a provincial tour !

" At the last meeting of committee I suggested that if we were anxious to show the Government all that could or might be done for the advancement of science, we ought, in mentioning what has been done by the B. A., to point out the great desiderata which for want of money we could not accomplish. I feel that this is due to the B. A. . . . We, the B. A., really were started into existence on account of the lethargy and want of general spirit in the R. S.; and if the latter is to do real good, they must imitate us or let us do the good work."

To J. D. Forbes, 16th March.—"You would see by the circular received from the Secretary that the move which I made in favour of a broad and liberal view of the Government grant to science prevailed, and that the principles acted on by the British Association have been adopted in the end. I say this, because I may assure you that it required all my energy to get the point carried. At the meeting of the Council where all was settled, I presided; and I hope you will approve of our losing no time and going to work *instanter* in employing all the money we can get in a good cause."

As the summer wore on, symptoms of the maladies of the previous year began to show themselves, so that a retreat

to the Continent was resolved upon. Our two invalids of
Buxton and Brighton repaired first of all to Vichy to try the
waters there. Murchison no sooner found himself among
the rocks again than his old geological ardour began to
revive. In a week or ten days he had explored the rocks
all round the watering-place, and collected materials enough
regarding them to furnish a paper afterwards for the
Geological Society.[1]

From Vichy he made a lengthened excursion into the
volcanic regions of Auvergne, which he visited with Lyell
twenty-two years before. Recent rambles among the extinct
craters of Italy had awakened in him stronger interest
in volcanic geology. Under the guidance of worthy M. le
Coq, of Clermont, he saw the ground again, and formed very
decided, though probably erroneous opinions, regarding the
succession of events in the volcanic history of that part of
France. In particular, we find him discarding the undoubtedly correct views as to the formation of the valleys,
which Lyell and he had adopted after Scrope, and returning to the true-blue convulsionist faith. "The denuda-
tions between these plateaux of basalt," he remarks, " could
never be explained by actual causes. Millions of years of
such a puny stream as the Dordogne could never deepen a
valley. All that actual causes effect here is gradually to
widen the valley and to fill it up." And he speculates that
subterranean commotion and violent upheaval at a point
from which the ground was starred by radiating rents, had
been the origin of the existing valleys !

[1] " The Slaty Rocks of Sichon shown to be of Carboniferous Age."—
Quart. Journ. Geol. Soc., vii. p. 13. Another paper was made out of the
notes of this Vichy sojourn—" On the Origin of the Mineral Springs of
Vichy."—*Op. cit.* p. 76.

GEORGE POULETT SCROPE, F.R.S.

This year the British Association was to meet at Edinburgh. In a letter of 31st May, Murchison had written to Hugh Miller about that gathering as follows :—" At a meeting of the Council of the British Association, held yesterday, I was named President of Section C (or Geology and Geography), with yourself and James Nicol as Secretaries for our special science, and with A. Keith Johnston for geography. As I moved that you be placed in this office, and as my motion passed unanimously (and indeed with acclamation in a full meeting), I trust you will not allow anything to prevent your accepting it. I honestly confess that no honour could be more gratifying to me than to occupy the Geological Chair in my native country, and if I know that the author of *The Old Red Sandstone* will be one of the Secretaries I shall be still more proud ; for I consider that *we* come from the same nook of land; the Black Isle and Cromarty being inseparable.

" I have written fully to the excellent and venerable Professor Jameson, telling him how, when the subject was first broached to me by my geological friends, I always insisted on declining the honour if every due respect were not paid to Jameson.

" I am writing by this post to Nicol at Cork. My great object is to revive and give greater breadth to the operations of Section C. I wish good and large subjects for discussion to be brought before us. Pray try your skilful hand at one of these, and get up also through others a good scene or two for our meeting-room. Sedgwick talks of dashing off a general memoir on the south of Scotland. He will never write it, but we shall perchance have it well spoken." . . .

Quitting, therefore, the *cheires* of Auvergne and the water-drinkers of Vichy, Murchison repaired to the northern capital to preside over the Section of the geologists. Always glad to have some striking announcement to make to his friends of the Association, he had the good fortune on this occasion to be able to publish a lucky discovery which he had stumbled upon while at Vichy. Bounding the eastern margin of the broad plain of the Limagne, the hills of the Forez stretch in undulating masses between the valleys of the Allier and Loire. French geologists, finding the rocks of these hills to be altered and crystalline, had placed them low down in the geological scale, a position, indeed, to which, judged only by mineral evidence, they might well be assigned. Murchison, in one of his first rambles, discovered a few fossils in these same crystalline masses, and in subsequent searching found more, which on examination proved to belong to the same great series as the Carboniferous Limestone of England. He at once saw the value of this fact in relation to the geology of the older rocks of central France, and in a brief notice to the British Association drew attention to it and to the corroborative evidence which it furnished, of the truth of a statement long before made by Sedgwick and himself, that some, at least, of the coal-fields of the Continent were laid down after the older Carboniferous rocks had been fractured and upheaved.[1]

At the close of the meeting Sedgwick and Murchison once more, and for the last time, became fellow-travellers. Leaving Edinburgh on the 9th of August, they first halted at Dunblane, not on geological errand bent, but to see the

[1] *Brit. Assoc. Rep.*, 1850, Trans. Sect., p. 96, and the same paper published in full in *Quart. Journ. Geol. Soc.*, vol. v. p. 76.

Cathedral, and ramble out to Sherriff Muir to look at the field of battle, and search for the stone from which the piper blew the war-pibroch that led Murchison's clansmen to their death. A little geology was done as the two comrades climbed together to the top of Ben Lomond. Thence they found their way to Inveraray, on a visit to the Duke and Duchess of Argyll. Of this time Murchison writes :—" We were truly received with open arms by the accomplished pair. It seems to have been a marriage made in heaven. Sound thinking, quickness, method, and a deep sense of religion combine to render the Duke a very remarkable young man. The Duchess is quite equal to him. With the good-humoured, placidly-jocose, and solid historian, Prescott, full of good manners and fresh stories told in a very easy and agreeable way, we had for two days the erratic fire of Sedgwick, whose episodes, flights, and parentheses fairly marvelled our Transatlantic kinsman. They naturally, however, delighted each other, whilst the steady *rôle* of C. Howard, and my occasional hammerings by the way, made up a very efficient table-talk."

After a few days of mingled talk, geology, and sport at Inveraray, our traveller reached " the hospitable retreat of Lord Murray, at Strachur." There he enjoyed more chat, saw some more geology, and helped to catch skate, crabs, and other treasures of the deep in the shore nets, and left with regret a home brightened by culture, music, and the intercourse of pleasant guests, drawn together by a genial and hearty hospitality. Several weeks were spent in these desultory employments, during which Murchison passed through a good part of the central Highlands, staying, as of old, with his hospitable friend Lord Breadalbane, and enjoying the sport to be had about the Black Mount and Glenorchy.

This Highland ramble, though not ostensibly for geological purposes, has an interest in the progress of the development of British palæozoic geology. Murchison had not looked in detail at any of the crystalline rocks of the Highlands since the early days when Sedgwick and he were companions among them. In the interval his experience had grown, and he now could look upon them with different eyes. Instead of being content to treat them merely as fragments of the primeval crust of the earth, he now traces in them everywhere proofs that they were originally only sand, gravel, and mud, which had subsequently been altered. He tries to make out their foldings and their order, conjectures that some of them may be of Silurian age, and though not able to ascertain much of precision and value, his notes and recollections of these autumn observations served him in good stead in later years, when he accomplished the last great feat of his life,—the establishment of the base of the rocks of Britain, and the structure of the north-west Highlands of Scotland. To this feat we shall return in a later chapter.

His subsequent wanderings in Scotland, after the hospitalities of the Highlands, were mainly geological, and are thus described by himself:—

"*13th Oct.* 1850.—MY DEAR SEDGWICK,—On my way through town a week ago I got the copy of your discourse, for which I thank you much. It is brimful of good stuff, and is to me a very gratifying *souvenir* of your powers.

"When I came out of the Highlands (after killing a stag at Glenorchy Forest, Lord Breadalbane's), I wrote to James

Nicol,[1] who was still at Edinburgh, and who had been making clear sections of the Lammermuirs, to come and join me at Glasgow to hunt up all the 'Siluriana' of the west coast, which he had never seen, and of which you had given such tempting accounts.[2] After a trip on the Clyde, to look at the clay-slates there, we journeyed to Girvan, in the neighbourhood of which we worked right hard for a week. What with our own labours in the field, and Sandy the fiddler's fossils, I was enabled to form some idea of top and bottom.[3] Nowhere have I seen such 'shelly sandstones' since I first explored them in the valley of the Meifod, or on the slopes of May Hill, etc. That they should have remained so long unknown is a great opprobrium to Scotch geologists. I find that the true *Calymene Blumenbachii* occurs in quantity low in the series of Girvan. The Government surveyors tell me that it is also the case in North ·Wales, where that same species goes down to the base, or very nearly, of the whole concern.

" I suppose you made out the relation of the conglomerates (what grand conglomerates!) to the fossil-bearing strata. There are several of these conglomerates, all of which were evidently taken by Macculloch for ' Old Red,' hence his erroneous mapping. On quitting the Stincher and Ballantrae, I

---

[1] Mr. Nicol had written some of the earliest descriptions of the Silurian rocks of Peeblesshire and the south of Scotland, as well as an excellent *Guide to the Geology of Scotland.* He had been chosen by the Geological Society of London to be its Assistant Secretary. He is now, and has been for many years, Professor of Natural History in Aberdeen University.

[2] Sedgwick had given a luminous account of the rocks of the southern uplands of Scotland to the British Association at Edinburgh.—See *Brit. Assoc. Report for* 1850. In Murchison's subsequently published narrative of his trip, he stated that he had received every encouragement from Sedgwick to examine the Scottish ground.

[3] Sandy Maccallum, a noted fiddler and fossil-finder at Girvan.

travelled southwards to Loch Ryan, and thence by Glenluce
to New Galloway, across the granite, and thence to Castle
Douglas and the headlands south of Kirkcudbright. At
Dumfries we got hold of Mr. Harkness, a very energetic and
hard-working young man, who is going to give us a paper of
some details of the county of Dumfries, and all its zones or
zone of Graptolites. It will be read early this session. At
Moffat we worked closely and hard for some days to find out
the mineral axis of the region. I believe that there, as elsewhere, particularly in the Lammermuirs, there are several
axes.

"I may give a little something to the Geological, for I
see that many *mémoires pour servir* must come out, and
many hands, hammers, and eyes must be exercised before
the south of Scotland can be brought into anything like
order. All the northern frontier of these rocks, laid down
by Macculloch, must be done over again.

"Tell me what you are going to do in this matter. Accept
of my wife's kind regards. Do not kill yourself with the
University Commission.—Ever yours, my dear Sedgwick,

"Rod. I. Murchison."

In a letter to Hugh Miller, Murchison expresses a very
sagacious inference, which his recent tour had led him to draw,
and which subsequent detailed examination has confirmed,
while at the same time he unwittingly utters a sort of pro-
phecy as to his final geological work, which really came in
the end to be fulfilled :—" As far as my researches go, they
teach me that whilst the Silurian rocks occupy so very
large a portion of the south of Scotland, they are far from
being very thick. Their apparent enormous development

is due to countless flexures.. . . . Unquestionably there is
no top or bottom to the series in the southernmost of the
Scottish Silurians. I look, therefore, for the top and bottom
of my system as it is now expanded in an European sense
in the Highlands of Scotland ; and so, after all my labours
and peregrinations, I think of returning to work at home,
and ending my days where I began them." . . .

The results of this rapid tour in the south of Scotland
were of considerable importance in relation to the history of
British palæozoic geology. The chain of undulating uplands
which stretches from Port Patrick to St. Abb's Head had
formed the ground of some of the most interesting observations of Hutton, Playfair, and Hall regarding the former
revolutions of the earth. From the time of these observers
it had been known that the rocks of those tracts, called then
by them " primitive schistus," were crumpled and twisted by
some of the early movements of the earth's crust, which preceded the deposition of the flanking sheets of red sandstone.
Hall, in one of his rides, had found fossil shells in them, and
the rocks were afterwards classed in the vaguely-defined
"Transition series." In later years Sedgwick had compared
them with his ,Cumbrian and Cambrian masses, but from
mineral characters rather than from fossil contents. Mr.
J. Carrick Moore and Professor James Nicol had afterwards obtained fossils, which placed the rocks definitely
in the Silurian system. Mr. Nicol had given much attention, especially to the eastern parts of the uplands, but
hitherto no clear views had been published as to the general
structure of the region, and still less as to any possible succession of the rocks determinable by fossils, though a singularly able and suggestive sketch had been given at the recent

Edinburgh meeting of the British Association by Sedgwick. Hence, following up the clue given by the Woodwardian Professor, Murchison, with his companion, Mr. Nicol, was able to show, by means of fossils, that well-marked representatives of his Llandeilo and Caradoc formations occurred in the south of Scotland,—a notable step in the progress of the extension of his Silurian domain over Britain.

Before, however, the notes of this Scottish ramble could be elaborated into the memoir, which in the following February was read to the Geological Society,[1] Murchison's restless industry had urged him into the midst of a new and fresh field of work. We have seen how much for a year or two past, ever, indeed, since his first Scandinavian trip, his mind had been exercised with questions of glaciers, icebergs, and débâcles. Each fresh mass of loose gravels and detritus met with in the course of his wanderings set him thinking anew on this subject, and the more he thought the more did he seem to feel the difficulty of seeing how solid ice could ever have been spread over the land, and could have produced, or have aided in producing, the piles of rubbish known generally by the name of "Drift." In Russia, Sweden, Denmark, the Alps, Italy, and lastly in Scotland, he had been brought face to face with these problems, the result being to make him a firmer believer than ever in great torrents and rushes of water let loose by subterranean upheaval. As soon as he had returned from Galloway, with its abundant mounds of ice-born detritus, he set to work upon the superficial deposits of the south-east of England. At intervals during the winter and spring he scoured the valley of the Weald and

---

[1] On the Silurian Rocks of the South of Scotland.—*Quart. Journ. Geol. Soc.* vii. (1851) p. 137.

the uplands of the North and South Downs, searching, as he said, for "Drift, drift, drift wherever I could find it." He communicated, as usual, his results to the Geological Society, —this time in a voluminous memoir specially devoted to a statement of evidence which seemed to him conclusive in favour of other and far more powerful action than any which the geologist sees in operation now.

In this memoir a protest is sounded against the views of those who, following out the doctrine of Hutton, refused to admit, as valid geological causes, kinds of agency different from those at present at' work. These writers, it declared, "have, in their too great eagerness to explain much that is still obscure, forced the former energy of nature into a quietude which is inconsistent with the proofs of her violent revolutions." But these "proofs" were precisely the same as those relied on by the author's opponents, who denied that they necessarily, or even probably, should be referred to violent and sudden action. Murchison's contention in favour of "great oscillations and ruptures" of the earth's crust lead- ing to the sudden breaking-up and submergence of tracts of land partakes of the very same vagueness which Playfair had long before complained of when arguing against the similar views of Pallas and De Saussure. He does not, any more than any of his predecessors, explain the *modus operandi* of his ruptures and débâcles. Time, that great arbiter of dis- putes, has decided against him in this matter of the origin of the "drifts" of the south-east of England, which, instead of being referred to violent convulsions and floods, are now regarded as among the most striking proofs of the long con- tinuance of quiescent sub-aërial waste such as is still going on. Nevertheless, in protesting against the doctrine that

nature's operations have always been, on the whole, as gentle
as they are at present, Murchison was probably in the right.
The interval during which man has been observing what
takes place around him has been infinitely too short to
warrant him in dogmatically assuming it to be the type by
which all past energy is to be measured.   From a wider
view of the history of our planet, the conviction arises that
as the sum of potential energy within the earth must have
been gradually diminishing as the globe has cooled, the
manifestations of subterranean action may have been lessen-
ing likewise, and may therefore have been far greater in
their earlier than in their later stages.[1]

Murchison's mind at this time seems to have been fuller
than ever of zeal for the former potency of Nature's operations.   Besides the memoir just referred to, he gave an
evening discourse to the Royal Institution·" On the former
Changes of the Alps" (March 7), wherein he returned to
the charge against the " uniformitarians," appealing to the
undoubtedly stupendous overthrows of the rocks of the
Alps as triumphant proofs that the present quiet uniformity
of nature has not lasted always, but has been disturbed by
enormous convulsions, which, in upheaving mountains, have
strewn their sides and the surface of the plains with vast
piles of shattered detritus.[2]

[1] See on this subject Jukes' *Manual of Geology*, 3d edit. p. 445, and
Quarterly Review, 1868, p. 204.

[2] The Friday evening discourses of the Royal Institution are usually
printed only in abstract, so that much of their detail is not preserved.
Professor Ramsay adds the subjoined note respecting this particular discourse :—" I was present at that lecture.   Murchison there spoke of
past epochs, and, among other things, declared that during the Carboniferous epoch the crust of the earth was rifted, and the heat from the
melted matter below, coming through the cracks, was one of the causes
of the tropical character of the Carboniferous flora ! "

The same creed of this *laudator temporis acti* appears still more strongly expressed in his private correspondence. For example, in writing to Mr. J. P. Martin, who had given some attention to the Drift, he says:—"You know my object in thus stepping forward to generalize the facts boldly, for a stand must be made against the fashionable nonsense of washing the Weald for thousands of years by ordinary sea-action." In a later note he remarks to the same correspondent :—" I have repeatedly shown in other works that operations of great violence, not of Lyell's quietude, have been repeated on and along the same axes or habitual lines of disturbance. Of these disturbances I only [now] deal with those of the last great geological revolution, in which the mammoths and their allies were specially massacred and destroyed." Again, in writing to Sir Charles Lyell with reference to the finding of supposed footprints of air-breathing animals low down in the older rocks of Canada—a position in which, according to his views of the progress of creation, they ought not to occur —he says, " I thank you for your discourse, and am not frightened out of my views by your pithy *P.S.* on Logan's footprints.[1] Hitherto, we know absolutely nothing of land animals in the Silurian world, and now that we find tracks of them, the animals are of the lowest or reptilian order. This is just as it should be.˙ If Logan had found the print of an aldermanic Robinson Crusoe's foot, as he was intent on realizing the first turtle-soup, then I would knock under. It is well we have some of these points to wrangle about, or we should become as quiescent as one of your most tranquil periods, and you would yourself call out for a revolution."

[1] See *Quart. Journ. Geol. Soc.*, vii. p. lxxv. The chelonian origin of these tracks was soon exploded. They were admitted not to be the work of vertebrated animals. See *op. cit.* viii. 223.

In a later letter to Sedgwick, these curious tracks from the Canadian rocks are referred to in connexion with the discovery of a remarkable fossil lizard-like skeleton from the yellow sandstones of Elgin, then and long subsequently considered to belong to the Old Red Sandstone, but now referred to the New Red Sandstone, or Trias :—" I have just been seeing the confounded frog that leaped in the primeval Devonian days of our whitish yellow siliceous sandstone of Burg Head and Spynie Loch, near Elgin. It is as beautiful a vertebrated little monster as any one of the Brown Coal or Œningen period. And he is to wag his tail next meeting, to the infinite delight of Lyell, who is inebriate with joy, and who will have him out in a new edition before we can launch him in our own Journal. Mantell describes it, and really has great merit in having suggested long ago that certain queer little gemmules of the Lower Old Red Sandstone of Forfar, long ago figured by Lyell, were the spawn or eggs of batrachians.[1] We have now got the Triton who laid them. At first I suggested that this siliceous sandstone might be of the same age as my Braambury Hill rock (Brora), which it much resembles ; the more so, as near this very Spynie Loch there are strata of the Oolite age ; but in the very same rock are the scales of the *Stagonolepis* fish of the true Old Red.[2]

"I hold (and I said so when Logan's tortoise-marks from the Lower Silurian of Canada were produced) that such proofs of the inhabitants of the land and fresh water of

---

[1] Reference is made to the fossil known as *Parka decipiens*, which is now recognised to have been the egg-packets of crustaceans.

[2] The *Stagonolepis*, formerly considered to be a fish, is now known to have been a crocodilian reptile. No true reptilian remains have yet been met with in the Old Red Sandstone, for the strata yielding the scutes of *Stagonolepis* are now referred to the Trias.

those early days can have no influence in changing our
general argument founded on marine succession. They have
found also the trail of a tortoise and some small plants—
potamogeton," etc.

In the month of July 1851, the British Association met
at Ipswich. Murchison says, "At this meeting[1] I induced
my associates of the Association to fill up the vacant letter E
in our list, unrepresented by any Section since the retirement of the medical men to their own Association, and I
constituted it as the Geographical and Ethnological section.
As long as they were subordinate to the geologists (Section C), the geographers were submerged."

At the close of the meeting he started for the south-
west of England to visit his uncle, General Mackenzie, at
Cheltenham. The veteran rejoiced to go through the old
campaigns again with his nephew, who somewhat briefly
chronicles the event as three days "with my uncle and the
' olim meminisse.' " In truth the geologist's thoughts were
now bending to a new scientific venture, and every day
was grudged which kept him away from the field.

---

[1] This is not strictly accurate. It was at the previous or Edinburgh
meeting that the rule passed, "That the subject of Geography be
separated from Geology, and combined with Ethnology, to constitute a
separate Section, under the title of the Geographical and Ethnological
Section." But probably Murchison meant to say that it was at the Ipswich meeting that the rule first came into operation. He was chosen
then as the first President of the new section.

CHAPTER XXI.

THIRTEEN years had now passed since the publication of *The Silurian System*. In this interval great strides had been made in the extension of the classification propounded in that work to the older fossiliferous deposits in many widely-separated parts of the world. The order brought out of the original Silurian rocks had proved to be no mere local phenomenon. The general type of life yielded by these rocks had been identified as characterizing the oldest fossiliferous formations all over the globe. Murchison himself had done much in carrying out this extension in Europe by his travels in Russia and in Scandinavia. He had likewise watched with growing interest its progress in the other Continents, and had kept himself abreast of the advancing knowledge. At length the idea occurred to him to gather up the gist of all this accumulating research, and present it to the world in a compact volume. His intentions, after he had made a little progress in the undertaking, were thus stated to his friend Hugh Miller, whose powers of graphic description he lost no opportunity of eulogizing :—

"I am preparing what I fear I shall fail in producing in a satisfactory form, a general and more or less popular view of primeval geology ; more particularly as respects all my Silurian foundation-stones.   I should not perhaps have thus endeavoured to put my house in order, which I thought I had sufficiently done in the first chapters of the work on Russia (where everything was described, as I shall again do it, from the beginning), had I not been urged to do so by my best friends, and particularly by Sir H. De la Beche and the Government geologists of my country. . . . I have been stimulated to get up a readable and general view, which I hope to be able to send you by next summer." . . .

For the next three years, Murchison's journeys were mainly devoted to the congenial task of gathering from fresh observations in the field, from comparisons of museums, from conversations with other observers as well as from their published writings, materials to be worked up into the new book which was to represent the actual state of Silurian geology.

He could not begin this new venture better than by visiting some of his old haunts along the Welsh and English border, and extending thence his examination into Wales, coming, as he did, with eyes of greater experience to see again his typical sections, and compare them with those in other parts of Europe.   At the end of his work, in which he had asked his companion of the previous year, Mr. Nicol, to join him, he thus reports progress to Mr. Ramsay :—

" *Holyhead, August* 15, 1851.—I cannot quit the shores of the bare but once well-clad dark isle of the Druids without thanking you for the loan of your maps, and for your introduction to that hearty good man, Mr. Williams of Llanfayringhornwy,[1] with whom we passed two days before

---

[1] The Rev. Mr. Williams was Professor Ramsay's father-in-law.

we came hither.   I have made a good and honest transverse section from Ludlow, by Welshpool, Meifod, the Berwyns, Bala, and thence, with deflections to Cader Idris, by Barmouth, Harlech, Tan-y-Bwlch, and Llanberis, to this place.   I am delighted with all your mapping, and above all pleased with Selwyn, who joined us at Bala, and took us in hand for Cader Idris.   It is a glorious region for a final base-line of all that is fossiliferous."

From Anglesea, where the section was run out to sea, it was natural to look to Ireland for some continuation of it.   Accordingly a descent was made by Murchison and his companion upon the Green Isle for the special purpose of looking after its Silurian sections.   The grievous destitution then prevalent in that country finds nearly as much comment in the letters and notes of the geologist as the rocks and fossils.   The subjoined extracts will show how the subjects were commingled :—

"I have been looking for many a long day for any base of Siluria in Britain, and until yesterday, between Blackwater and Talliconner Bridges, I never saw it, as made up of fragments of mica-schist, quartz-rock, etc. ; in short, of all antecedents.   To-day we had no reason to hope for a similar base-line.   But Griffith's map gave us an outline, and his notes a direct indication of the fossil-beds, *i.e.* of Silurian, related to primary rock ; and in following it, despite of wind and rain, we worked out all our points north and north-west of this comfortable little inn, and the table is now strewed with the relics of the lower fossil group."

" A penny given in alms brings out a colony of beggars. They seem to rise from the earth quicker than mushrooms after a genial shower, and in a country where no man seems

SIR RICHARD GRIFFITH, Bart., F.R.S.
From a Photograph by Maull.

to live. In truth, the cabins, such as are left, and the blocks
and boulders so resemble each other, that you cannot tell
where the people are."

"*Galway.*—The export is *nil*, beyond Paddies and the
stones that would not feed them. . . . Roofless houses
and pompous fat sleek Papist priests are disheartening to
see."

"Passed the nick or opening in the mountains where
the tradition is that the devil bit a piece out, and flying
away with it, found it too heavy, and dropped it at Cashel.
If geologists had lived in the old days, they would have defeated these lying inventions, for the hill with the nick is
Old Red Sandstone, and the rock of Cashel dark grey Mountain Limestone.

"I have looked at some of the Connemara sections, as
well as those of Dublin, Wicklow, Wexford, and Waterford,
and bring back to my mind all that I saw in former years.
. . . The multitude of intrusive granites, and the horrible
spread of drift and bog are intolerable obstacles."

Before starting on this tour he had asked Sedgwick for
any suggestions as to points requiring elucidation. When
he got back to England he wrote to the Cambridge Professor :—" I wish I could have so arranged matters as to
meet you in Scotland ; but, to say the truth, I had much
rather be in Ireland, despite of misery and rags. . . . I now
regret having left Ireland, and have some thoughts of returning to look up two or three typical sections this autumn, if
the weather prove dry and settled."

In spite of their disagreement on one geological question, these two friends continued to be on intimate and
cordial terms. This very summer, for example, Murchison

could still write as follows to his comrade :—"J— being a most pugnacious and vindictive Frenchman, judges others by his own measure, and thinks that because you and I have had a wrangle about Sil. and Cam. we are, or may have been, estranged ! He little knows the nature of our friendship." We may take another sample of their friendliness, and, at the same time, a whimsical picture of the Cambridge geologist, from a letter of Murchison to Whewell, of the previous year :—

" I learn by the papers of this day that Sedgwick has had a very bad accident, and has fractured his right arm, and as, of course, he is in durance and suffering, I do not think of writing *to* him. I will, however, be very much obliged to you to convey to him my kindest regards, with my hearty wishes that the year which we enter on to-day may bring very different fortune to him than that which has passed. In the summer he wrote to describe the abscess in his leg as a ' volcanic eruption.' Dislocations or faults in various parts of his body are, poor fellow, no new things to him ! He is specially unlucky on horseback. I have just been out shooting [at Up Park], and young Tom Erskine, one of your lads at Trinity, who was with me, tells me that Sedgwick's horse is called *The Mammoth,* and is an enormous animal ! No wonder, then, if he is bruised so much by such a catastrophe as being rolled under the monster.

" It is high time, however, that my dear old friend should abstain from these gambols, and if he can be persuaded that the gout can be as well kept off by good peripatetic discipline as by rolling over on horses and mammoths upon the hard road (I believe even ' Brown Stout ' tumbled down with him), I hope he may never mount again."

Indifferent health and numerous ailments had undoubtedly done much to hinder the work of the Woodwardian Professor. He took his revenge by playfully satirizing them and their victim. Did ever man, for instance, draw a more absurd picture of himself than ·is given in the following letter, written (14th April 1851) from his canonry at Norwich? In estimating the amount of Sedgwick's work, and the hindrances he had to struggle against, we must bear in mind his continual conflict with bodily infirmity, which even though sometimes partly imaginary, was none the less irksome.

" MY DEAR MURCHISON,—I send you a cheque to be put to the subscription account. I wish I could send more; but I am picked to the bones, and though I have a good income for an old bachelor, I do not always contrive to make the ends meet, so that I am poorer and poorer every year. But I do mean to mend my ways, turn a churl, and save money for my executors.

" I have had a miserable residence. The influenza never quite left its hold upon me, and the cold cathedral confirmed its grasp. I did my best to barricade my lungs from the cold air. Ever since I came into residence I have gone to my stall with a black velvet cap arching over from the *occiput* to the *os frontis*, and a tremendous black respirator over my jaws and muzzle. So that I look like a true angelic church-militant going to war with a black helmet and a visor so far uplifted as just to show a running snout and a big pair of blear eyes. My cold *is* better, though my lungs are still impatient of cold air.

" But as the cold went out the gout came in. So, to complete my costume, I now wear a tremendous pair of black snow-boots over my shoes and ankles. Such is my daily costume at the altar-table, where I sit and stand in state; and such a sight was never before seen at Norwich. So much for the outer man.

" My spirits are all gone, my memory is shattered, and my temper is turned to distemper. All the bad parts of the old Adam are vigorously thriving, and whatever good there might once have been in him is all gone, without leaving so much as a *caput mortuum*. But all things have their end, and so must this true history. I bless my stars that I have still the power of grumbling. Everything is out of order.

" My kindest remembrances and good wishes to Lady Murchison. Through cold, wet (and it has rained every day since I came hither), bronchitis, catarrh, gout, and hyp., ever yours,   A. SEDGWICK."

A few extracts from Murchison's diary of this autumn will further illustrate the odd blending of science, shooting, and social enjoyment in which his life was spent at this time :—

" *Keele, Staffordshire, Sept.* 4.—Yesterday, when I left for Sneyd's Place, old Lord Combermere (the great Mogul's ' Son of the State,' ' Sword of the Empire,' etc.) would accompany me to Audlem Mill, insisting that there was a rock there. He rode his famous pony Thumbscrew, and I was in my Lady's open phaeton. The pony, smelling some fresh hay at a low stable door, rushed in, and I thought the old Viscount's back would have been broken. The chaos in the little stable was terrific, the pony and my Lord rolling on the straw, and a huge miller's cart-horse lashing out with his hind-legs over them. The agent and self pulled the old

chief out by his legs,—trousers all torn. It was a great
escape, and he galloped home, cheering as he went."

"From Downham, where I had some indifferent partridge-
shooting, I passed on to Cambridge to study the advance
made in the Woodwardian Museum by the labours of Mr.
Frederick M'Coy, a clever young Irishman employed by
Sedgwick."

"*Sept.* 17-20.—At Broadlands. Partridge-shooting, diplo-
macy, and fun in a charming place.

"*21st.*—At Nursted, and on Sunday went to church at
Buriton with my wife, thirty-six years after we were married
in it."

"*Havant,* where I write this. Forty-two years last
January since I marched through here with the 36th from
Portsmouth, after Corunna. Horrible Sunday train; pleas-
ing only in exhibiting so many pretty country lasses.

"*22d.*—At one o'clock this day buried my poor sister,
Fanny White, in the new cemetery at Tunbridge Wells.
This funeral-day not only brought back the boyish days of
Durham, when I, a boy of six, was welcomed by a fine
exuberant elder half-sister, but drew tears to my eyes, and
led me to pass in review her chequered Eastern life, and the
many many trials she had passed through. Then came
before me the recollection of her music, and with what a
touch and feeling did she make us all spring up to dance a
Highland reel! We interred her on a lovely, mild, and
glowing autumnal morning, in the new cemetery, with an
exposure to the rising sun, which was thus a true emblem
of her hopes, as founded on a spotless life and a firm faith!
She ever spoke kindly of all persons, and cannot have left
one enemy behind."

"*Shoreham.*—Halted two hours to examine the clay brick-earth pits or diggings north of the town. Eight to ten feet of unlaminated yellow clay exposed, with some small pieces of chalk and flint, etc."

"*Binfield Park.*—Flint detritus, tertiary sands, grey-wethers, etc. Here is a good quotation to apply to the dull geology of the Wealds of Kent, Sussex, and Surrey, and the heavy inhabitants thereof :—

'Old Andred's Weald at length doth take her time to tell
The changes of the earth that since her youth befell!'
(*Drayton's ' Polyolbion.'*)

"Following the Drift westwards from Hants into Wiltshire, here I am, with my old fox-hunting friend, Thomas Asshton Smith,[1] to see the opening of the fox-hunting session. . . . The fox crossed the Avon and threw the whole field : the body of hounds close at her, raced for four miles, —a brilliant burst.[2] The vistas on these sloping hills, never too steep, are glorious—the finest. country in the world for a gallop. . . . S. was a naughty boy, as his mamma said, when he was three years old, and his papa whipped him; whereon the young squire, as he told me, resolved to set his mind against all control, and he has had his own way ever since (*æt. suæ*, 76). If he had not been flogged he thinks he might have been a different man. He is now a hydropathist, having been a homœopathist; whereon Dr. Quain, being railed at at a dinner-party thus :

[1] This was one of the most noted sporting and yacht-building men of his day, proprietor of the great Llanberis slate-quarries, and one whose acquaintance with Murchison dated from the old fox-hunting days described in Chapter VI.

[2] "This, I think," says Professor Ramsay, "was Sir R.'s last mount. Mr. Asshton Smith mounted him, and he afterwards told me he was so shaken he would hunt no more."

'Well, doctor, what do you say now to Mr. S. having ratted?' replied, 'Alas! alas! and the worst is, he is a water-rat!'"

After a prolonged series of autumnal visits to country friends, Murchison, late in the year, once more took his place amid the work and bustle of London life.

He had now again become President of the Geographical Society, and was giving a great deal of time and thought to the duties of that office, especially to the most efficacious means of increasing the prosperity of the Society. His forthcoming work on Siluria made, of course, but slow progress; and, indeed, his attentions to the younger Society were such as to rouse a little good-humoured jealousy in the minds of some of his older associates among the geologists. He excuses himself thus to the Master of Trinity :—" I have been too geographical, but was forced into the position by a little flattery of my usefulness. The public men think much more highly of me for having been the first who worked out mentally the Australian gold, 1845-46, by comparison of what I called the Australian Cordillera with my auriferous Ural, and for dwelling on it in successive years until the diggers discovered it."

This is perhaps the most convenient opportunity for taking notice of Murchison's relation to the discovery of gold in Australia, on which, as is evident from the foregoing sentence, he prided himself not a little. Two years later, viz. in the winter of 1853-4, much controversy arose, both here and in New South Wales, as to the respective merits of different claimants. Murchison, who had previously shown considerable sensitiveness as to the due recognition of his own claims, again threw himself at that time into the controversy. His own view of his position may best be gathered from the subjoined letter to the *Times* :—

" To the Editor of the Times.

" SIR,—In commenting upon a recent vote of the Legislature of New South Wales, by which Mr. Hargraves was recompensed for having first opened out the gold-fields of that colony, your correspondent at Sydney, after an allusion to the inductions of science, has thus spoken of me :— ' Sir R. Murchison pointed out the similarity of the Blue Mountain chain of Australia to that of the Ural in 1844 ; it was considered a mere speculation, and, as to any practical effect, might as well have been written of the mountains of the moon.'

" As my relation to this subject is thus summarily settled, I must, for the credit of the science which I have so long cultivated, state the following facts :—The comparison above alluded to was drawn by me after an exploration of the Ural Mountains and an examination of rock specimens gathered from the whole eastern chain of Australia by my distinguished friend Count Strzelecki. In 1846 I renewed the subject, and applied my views practically by inciting the unemployed Cornish tin-miners to emigrate and dig for gold in Australia. Both of these notices were published (1844 and 1846), the one in the volumes of the Royal Geographical Society, the other in the Transactions of the Royal Geological Society of Cornwall. I have every reason to believe that they are the earliest printed documents relating to Australian gold ; and, unquestionably, they were both anterior to the discovery of the Californian gold. Let me further state that they produced results ; for in 1847 a

Mr. W. T. Smith, of Sydney, acquainted me that he had discovered specimens of gold, and a Mr. Phillips, of Adelaide (equally unknown to me), wrote to me announcing the same fact. It was also in the same year (1847) that the Rev. W. B. Clarke, whose geological labours have thrown so much light on the structure of New South Wales, published his first essay on the subject of gold in the *Sydney Herald*, and referred to my previous comparison with the Ural.

" Seeing, therefore, that I had become a sort of authority upon Australian gold, and that the metal had actually been discovered and could be profitably worked under due regulations, I addressed a letter in 1848 to Her Majesty's Secretary of State for the Colonies, explanatory of my views, urging the desireableness of such a geological survey of the region as would realize auriferous and other mineral products. That letter, written three years before the operations of Mr. Hargraves, has, through the courtesy of the Duke of Newcastle, been printed among the papers relating to Australian gold presented to both Houses of Parliament, August 16, 1853.

" My scientific friends are indeed well aware that on various occasions between 1844 and 1851 I addressed public meetings on the same important phenomenon ; and I should not have sought to encroach on your columns had not my name been associated in your widely-circulated journal with the mountains of the moon, of which, I regret to say, I have no knowledge, whether they be situated in the heart of Africa or in our nearest neighbour of the solar system.—I remain, Sir, yours very faithfully,

" RODERICK IMPEY MURCHISON.

" 4 CIRCUS, BATH, *Jan.* 10, 1854."

In spite of the frequent reference to "science" and "scientific induction" in the course of the controversy, it is not easy even for a partial friend to discover in what way Murchison's share in the finding of gold in Australia could be regarded as in any way scientific, or more than a lucky guess. He had come home full of his doings in the Ural Mountains, and with some rather crude notions as to the mode of occurrence of gold throughout the world.[1] At that time he met Count Strzelecki, and saw his maps and the collection of specimens which he had brought home from Australia. Ready to find analogies with his Urals, Murchison noted a general similarity of trend in the Australian and Russian mountain ranges. On looking at the specimens, he recognised many fragments of quartz, and when comparing the Australian with the Russian rocks, remarked that as yet (1844) the former had not yielded gold. He knew nothing personally, and very little more by report, of the geological structure of Australia. When, therefore, he advised the unemployed Cornish miners in 1846 to emigrate and dig for gold in Australia, he had absolutely no scientific grounds on which to base his advice. All he knew was that there were crystalline rocks with quartz veins in Australia as in the Urals. But the same might have been said of almost any country on the face of the earth.[2] His advice,

---

[1] One of his favourite, but singularly unphilosophical, notions on this subject was, that gold was the last-created metal, and only occurred therefore in the uppermost parts of any formation. Theoretically, according to this doctrine, there could be no profitable gold-mining by sinking shafts into the solid rock. He clung pertinaciously to this notion, until the successful reef-mining of Victoria compelled him to modify it.

[2] Even now, with all the experience of gold-mining since 1846, he would be an exceedingly bold geologist who, from the inspection of a few bits of quartz, none of them containing gold, should pronounce on the auriferous nature of the country whence they came. Science has not been able to

however, was, under the circumstances, as good as could
have been given, for if the miners found no gold, they at
least would be in a colony where other openings for
gaining a livelihood presented themselves much more
abundantly than at home.   They could hardly lose by
emigration ; they might gain a good deal.

When gold had once been actually found, it was
natural to desire a thorough examination of the country
yielding it. The world was ringing at the time with the
newly-discovered marvels of the Californian El Dorado, and
no one could tell whether a rival to that region might not
be found in Australia.   It was at least worth while to explore. In urging this matter upon the Government, therefore, Murchison showed an enlightened desire for the spread
of geological knowledge and industrial development.   But
he did not thereby establish any claim to have foretold on
sound scientific grounds the really auriferous character of
the Australian rocks.

Count Strzelecki appears to have been the first to
ascertain the actual existence of gold in Australia.   But at
the request of the Colonial authorities the discovery was
closely kept secret. The first explorer who proclaimed the
probable auriferous riches of Australia on true scientific
grounds—that is, by obtaining gold *in situ*, and tracing its
parent rocks through the country—was the Rev. W. B.
Clarke, M.A., F.G.S., who, originally a clergyman in England,

make so clear the circumstances which have determined the presence or
absence of gold in quartz.   If the geologist declares that the quartz will
prove auriferous, he has no more scientific ground for his assertion than
any empiric or miner with a divining-rod.   He makes a guess, and if the
prognostication should be fulfilled he may talk of his luck, but has no
cause to boast of his science.

has spent a long and laborious life in working out the geological structure of his adopted country—New South Wales. He found gold in 1841, and exhibited it to numerous members of the Legislature, declaring at the same time his belief in its abundance. While, therefore, geologists in Europe were guessing, he having actually found the precious metal, was tracing its occurrence far and near on the ground. It is only an act of justice to render this acknowledgment, which Murchison himself, through some over-estimate of his own contribution to the question, and probable ignorance of what had really been done by Mr. Clarke, never made.[1]

In the spring of 1852 Murchison is found taking infinite pains over the preparation of his address for the May anniversary of the Geographical Society. Seven long years had passed since, in quitting the President's chair, sanguine of the success of Franklin's expedition, then just sailing from our shores, he had wished it God-speed—seven long years of suspense and slowly dwindling hope, and of noble efforts for the succour of the lost. Against the ever-increasing conviction that further search for the missing explorers was useless, there were some who yet resolutely struggled, clinging to every feeble thread of evidence that might seem to warrant even the possibility of survival.

---

[1] Count Strzelecki's observations were published in 1845, in a volume entitled *A Physical Description of New South Wales,* containing a sketch-map and a good series of carefully drawn sections. The labours of Mr. Clarke have been the subject of minute inquiry by the Legislature of New South Wales (1861), and the result of the investigations of a Special Commission appointed for the purpose was to show that his services had never been adequately recognised. See *Report from the Select Committee on the Services of the Rev. W. B. Clarke,* ordered by the Legislative Assembly to be printed, 1861; also a pamphlet entitled the *Claims of the Rev. W. B. Clarke,* Sydney, 1860, where, in an Appendix, references are given to the dates and proofs of his services.

Foremost among these noble-hearted believers was Lady Franklin. At her own charges she had equipped one searching expedition, and had largely contributed to the outfit of two others. In the spring of 1851 two half-wrecked ships, perched on the ice, were said to have been seen drifting southwards along the shores of Newfoundland. Could these have been the ill-fated "Erebus" and "Terror"? The very possibility of such a fact sent a thrill of excitement through the people of England, and gave a new impulse to the desire either to find and rescue the survivors, or at least to learn their fate and bring home their memorials. The President of the Geographical Society had interested himself keenly in the fitting out of the successive searching expeditions, and now in his address he places the subject of Arctic exploration and the fate of Franklin in the front rank of interest. He refuses to take the desponding view, which was now growing general, but, in the interests alike of philanthropy and of geographical discovery, rejoices in the prospect of renewed search.

Other topics of permanent interest find a place in the same elaborate address. The early labours of Livingstone in South Africa are alluded to, together with those of Mr. Galton, and the ingenious suggestion (verified three years afterwards by Livingstone)[1] is made from all the data then available, that Africa had originally had a basin shape, formed by an outer range of harder and higher rock-masses sinking into a vast and less elevated central area, and that this original structure will be found still in great

---

[1] See a reference to this hypothesis and its verification in Livingstone's Dedication to Murchison of his *Missionary Travels and Researches in South Africa*, 1857.

part maintained, whether the rivers escape through rents towards the sea, or flow inwards to lose themselves in lakes or sands. Proposed commercial routes across the American Continent, and from the Mediterranean to the Persian Gulf, and gold-fields all over the world, are discussed with the general and gratifying progress of geography.

In the preparation of such a detailed and voluminous document, the President necessarily depended a good deal upon the aid of the active Secretary of the Society, Dr. Norton Shaw; but, in his usual thorough and matter-of-fact way, he had done his best to make himself master of all the topics on which he had to touch. Indeed, geography and the Society were becoming each year more interesting to him. His soirées now partook largely of a geographical element. Every traveller of note who happened to come to London was sure to be seen at them, while, at the same time, the members of the Society mingled there with other men of science, literature, and art. In this way he strove to give a help to the *esprit de corps*, and at the same time bring the Society more prominently forward. The membership was steadily increasing, the funds, too, had considerably grown, and there were no debts.

In helping the advance of the Geographical Society, Murchison brought out in strong relief one of the most notable features of his life, often alluded to in the foregoing narrative, and still to receive further illustration in later pages. He possessed very considerable influence with leading men of all shades of politics. He met them continually in society; he asked them to his house, and was in turn invited to theirs; shot partridges with them in the country; and, having had a previous military and sporting life, was regarded by them as coming nearer to themselves in tastes and pursuits than the typical learned man of science, who was supposed to be able to talk only on his own pet subject,—beetles, chemistry, mathematics, or whatever that subject might be. The bundles of letters addressed to him, and still extant, furnish curious evidence as to the nature of his influence, the way it was used, and the multitude and variety of suitors for it. At one time the petitioner is an old friend, whose nephew, a most deserving youth, needs a helping hand towards getting a presentation to an Oxford Exhibition, in the gift, or otherwise within the influence, of a nobleman with whom Murchison is earnestly requested to intercede. Next comes a poor widow, whose great-grandfather, or other remote relative, had known some equally distant Highland ancestor of the geologist, and who would fain get her son into the Scottish Hospital, or other charitable institution in London. Then a Scottish professorship falls vacant, and instantly siege is laid to gain his influence and aid, which, if secured, are probably soon set in motion. Applications for testimonials seem to have been sometimes almost as plentiful as tradesmen's circulars ; nevertheless, in spite of other abundant calls on time and patience, he did his best for his clients, as their letters, grateful for aid and kindly sympathy, remain to witness.

The Cambro-Silurian fire which had been smouldering for a little, broke out with renewed and unexampled energy in the spring of 1852. Murchison had not been doing or saying anything fresh on the subject of North Wales. Indeed, he had made no material addition to the announcement of his geological addresses of 1842-43 already referred to. Meanwhile, however, the Geological Survey had been steadily unravelling the structure of North Wales, and had pronounced the rocks there to be in the main only the extension, in folded and contorted masses, of the Lower Silurian formations of Murchison's original Silurian region. The officers of the Survey restricted the term Cambrian to the thick unfossiliferous deposits which, in several areas of North Wales, were seen to form the base on which the fossil-bearing Lower Silurian rocks rested. Sedgwick, however, refused to accept this nomenclature. In a paper read to the Geological Society on 25th February, he gave forcible expression to his dissent, using language with respect to his old comrade, which, though probably far from being meant to offend, was yet felt by the friends of both antagonists to be too personal. Murchison's feelings are told by himself in a letter written to the Woodwardian Professor two days after the reading of the paper :—

" MY DEAR SEDGWICK,—In enclosing you one of my cards for soirées, let me beg of you to prepare the abstract of your paper, so that there should be nothing in it which can be construed into an expression on your part that *I* had acted unfairly by you. This is the only point which roused my feelings the other night, and made me speak more vehemently than I intended. But I did intend to tell the meeting in reference to that very point (what I forgot to say) that I have over and over urged you to bring all your fossils and complete the subject you had undertaken. It is no fault of *mine* that you did not do this. . . .

" But enough of this. I cannot presume to do more than speak frankly to you ; and whilst I daresay you will not change your opinions about nomenclature, I again entreat you to allow nothing to appear in print which can lead the world to suppose that we can quarrel about a name.

" I have sent Hopkins a most amusing letter of old Von Buch about the Drift and Erratics, which ought, I think, to be printed as an appendix or *P.S.* to the President's speech, and I have begged him to let you see it. So send it back to me, and pray let us wrangle no more about the *vexata quæstio.* We have done many a stroke of good work together, and if we had waited to describe the whole Principality and the bordering counties of England, the lamentable position in which we now stand would never have occurred. But I am told by Logan and others that if I had delayed a single year or two in bringing out my Silurian System with all its fossils, the Yankees would have anticipated me. And you well know that Wales, North and South, was not to be puzzled out in less than many years of hard labour.

" I have been grievously pained to be set in antagonism to you, but I can solemnly assure you that I know no possible way by which my present position could be altered without stultifying my original view of the Silurian System as a whole, and my confirmed and extended views respecting it as acquired from a general survey of the world.—Yours, my dear Sedgwick, most sincerely,

" Rod. I. Murchison."

The reading of Sedgwick's paper produced a lengthened debate and some commotion at Somerset House. But it nevertheless passed the review of the Council, and was printed and published in the Society's *Journal.* When, however, its pungent language, stripped of all the humour and *bonhommie* of oral delivery, came to be calmly read in print, there was a very general expression of sympathy with

Murchison.   At first the Council decided to cancel the printed part—a curious decision to make when the *Journal* had already been published and circulated over the world. Ultimately Murchison was allowed to write a short historical statement by way of reply, which was placed in the next number of the *Journal* immediately after Sedgwick's paper.

It was a most temperate and friendly rejoinder, showing the writer's very earnest desire to keep the peace, and persuade the world that, in spite of appearances, no personal quarrel existed between him and his old friend.   Later in the year he again writes to Sedgwick :—

"*Nov.* 22, 1852.— . . . I can safely aver that I did nothing whatever to induce the Government surveyors to adopt the line they have, and I never went into your region until they had quite settled all their nomenclature, except a skirmish to Snowdon in 1842.   It has a very bad effect upon the progress of our science to see Sedgwick and Murchison trotted out as controversialists.   All our oldest and best friends regret it sincerely, and the more so as there is really nothing in the philosophy of the case on which we differ.   We agree in the grand doctrine of a progression of creation, and we both start from the same point, now that the data are as fixed in the British Isles as they are in other countries.

" Why then can you not state *totidem verbis* that the fossiliferous part of your Cambrian is my Lower Silurian, or if there be this remarkable community of fossils between the upper and lower groups, why not call on your part the lower half, as exhibited in Britain, ' Cambro-Silurian,' a term you once proposed for what really now proves to be the same as Llandeilo ?   Such an explanation from you would let the world know that there was no philosophical dispute between us. Rely upon it that the more they are examined, the more will these two things be united, and I therefore wish that there should be nothing dissonant in our mutual expressions respecting them. . . .

" I did not intend to have said a word (when I began) on this topic ; nor will I ever write more on it. *Nomine mutatur*, the thing remains the same. If you are ever so gouty, and as you term yourself, stupified (which I do not accept), when you receive this, you must not quarrel with me for telling you all my thoughts and feelings. I have too sincere a regard for you not to do so, and the matter must now be in your own hands."

It is pleasant to turn from this sad part of the narrative to the amusing and most characteristic epistle of the veteran Von Buch, referred to in the last letter but one. Such sturdy adherence to the old belief in the midst of modern defections and heresies has something almost heroic about it. But in spite of its conservatism, the geological reader will recognise the shrewdness of observation and the wide range of knowledge, even in minute details, which helped to make Von Buch so deservedly honoured and admired as a magnate in science :—

" BERLIN, 22*d February* 1852.

" DEAR SIR,—Lolo is arrived.[1] He speaks very loudly to me, and tells me of many happy moments in the Tyrolean Valleys, at Inspruck, Pfunders, Eyers, at Meran, at Trento, and at Venise, where he would graciously accept some of Danielis biscuits. It is a masterpiece, and so it is hold for

---

[1] This was a dog which had been Lady Murchison's travelling companion in the Tyrolese and Italian tour. Von Buch took a great fancy to it, and used to carry biscuits and sweetmeats for it in his pocket at Venice. It had now been sent as a present to him.

by every one who sees him—Lovely beast—*Requiescat in pace.*

"Now have I seen and perused your most interesting Drift paper.   The whole was quite new for me.   I think it is a phenomenon belonging to the catastrophe of the separation of England from the Continent, so ably and convincing illustrated by M. Owen, after the quadrupeds creation and before the mans appearing.   That was after the dispersion of the great blocks of granite and gneuss; after what is called the *Glacial Drift,* the *Glacial Epoch!*   Angels and ministers of grace defend me!   *Glacial Drift!*   It was once a frightful disease in Switzerland, a kind of wide spreading cholera; it passed slightly over Germany, but went over and has fixed itself on the other side of the channel.   It will not attack the poor, but the man of the greatest genius, power of mind, and energy.   But every disease, every epidemic fever comes to an end—and— an Icy period, an icy floating between our time and the Tertiary world!!   Never.   There is not the least proof of such a supposition, and when able men will explain many curious facts by such means they are not aware of the Holy Scripture, which says, *Non, fingendum aut excogitandum.* The beds of Arctic shells shall prove an *Arctic climate!* Oh no!   far off.   There was no way for other shells to enlive the northern seas.   But since the opening of the Strait of Dover they could come from the Atlantic, and drive away Arctic animals.   The beds of Arctic shells (*Ostrea edulis, Buccinum undatum, Littorina littorea, Cardium edule*) are only found round the enclosed north sea, and not farther south. The last beds to the south are described by Mr. Sedgwick on Warden Cliff, Sheppey.   Nothing the like is to be seen to the other side of Dover Strait along Devonshire, Cornwall, or Normandie and Bretagne. So it was not the *climate*, but the impossibility of Atlantic shells to reach the shores of Scandinavia and Scotland, which have accumulated arctic shells where there is no more their prevailing abode.

"The shells in the Baltic are very small, and disappear entirely since the Gulf of Finland. The saltness of the sea-water is not great enough. So it was in the Drift time, and the opinion of a communication of the White Sea near Archangel with the Baltic in the Drift time is certainly *erroneous*. The shell beds are found in the interior of Holstein, 262 feet above the sea, and 70 miles from the North Sea, at a place called Tarbeck, a curious fact, which proves a canal from the North Sea to the Baltic in ancient times ; but these shells have no extension in the Baltic. The limit of the shell beds with large *Ostrea, Cardium, Buccinum* rises to the north of Fünen, then from the Kullen to Wermeland, Christiania, and along the Norwegian shore to the North Cape. But such beds are not found anywhere on the coast round the Baltic, much less between Petersburg and Archangel, where certainly they would have been deposited, if ever there had been a communication from sea to sea. Your discovery of the shell beds on the borders of the Dwina proves it to evidence. Even all the deposits between Stockholm and the Wenern described by Sir Charles Lyell are not formed of the large shells on the border of the North Sea, but only of the small shells as you will see them still living in the Baltic. So old was the separation of the inland sea from the North Sea ! *Pereat* the Arctic climate ! *Pereat* the Glacial Epoch! May the Geologists of Scratchings and Etchings and Fissures delight in such cool views. I will rather follow the lessons of the Geologists of crystallisation, or those of sediments.

"With the warmth of a Tertiary Epoch, I continue to be, dear Sir, your devoted servant and admirer,

"LEOPOLD VON BUCH."

The preparation of the new work on the Silurian System had all this while been making but slow progress, although part of it had already passed through the hands of the printers. In the spring of 1852 the author, always able to get on faster with his hammer than with his pen, took flight from the turmoil of London life and tried to get a little quiet and make some way at Buxton. From that retreat he sends (9th April) a letter to his friend Barrande, from which a few sentences may be quoted :—" The preparation of a long geographical discourse (which I send you a copy of) and multifarious London distractions (including the management of a large Society) necessarily checked my progress, but I have been getting on here, and hope to go to press in the autumn, and publish in the spring or early summer of next year. The *Silurian System,* now written upwards instead of downwards, as before, and *Russia in Europe* will form the 'stock-work' of the volume. I have taken all the figures of my best quarto plates of the *Silurian System* and reprinted them in octavo plates, so that, with many new woodcuts to illustrate new data, and an occasional glimpse of foreign analogies (at the head of which is your Bohemia), I hope to comprise the whole in one thick octavo.

" I hope to give in a practical way, and in comparatively few words, a knock-down blow to the theorists, who oppose all evidences of a beginning, and who deny a progression of creation. This, and the demand for a work which shall bring up the Silurian System of 1835 to what it is in 1852-3, are the real motives for my publication. The special object of this letter is to ask you if you wish me to say anything of you or your labours at the Belfast meeting of the British Association, which begins on the 1st September."

But though a sojourn in the country may be successful in putting an end for the time to the distractions of life in London, it does not place one beyond the reach of the post-bag. Murchison's correspondence, always large, involved this year an especial amount of work. By a curious combination of circumstances, a succession of Professorships fell vacant in different Colleges throughout the country, and he exerted himself most vigorously on behalf of friends of his own who were candidates. The amount of personal trouble he cheerfully undertook in some of these cases must have sadly interfered with his literary labours. For instance, a chair in one of the Scottish Universities had become vacant, and he determined, if possible, to have it filled by an old friend and fellow-traveller of his own, of whose abilities he had a high opinion. He besieged the Crown authorities in whose gift the appointment lay, and received a verbal promise of the chair for his friend. Months, however, passed, and no formal presentation was made; He again applied to his friends in the Government, but before anything further could be done, the Ministry of the day resigned. Nothing daunted, he successively laid his case before the new Premier, Home Secretary, and other members of the Government, and, after further provoking delay, carried his point and got his comrade appointed. Imme-

diately afterwards he entered into correspondence with the
Lord-Lieutenant of Ireland, and succeeded in securing the
presentation to two vacant Professorships in that country
for the candidates whose cause he took up. Among his
papers, too, there is a foolscap sheet of MS. in his own handwriting, the draft of a letter to the Premier, interceding for
an annuity from the Crown, as a mark of approbation for
the arduous and successful labours of his tried friend John
Phillips. That geologist, however, was immediately afterwards appointed Deputy Reader of Geology at Oxford.
Kindly, witty, able, and eloquent Buckland had succumbed
to the insidious malady which had slowly clouded his
faculties, and a few years later brought him to the grave.

In the midst of these exertions for others, Murchison
was summoned this summer to Oxford to receive the degree
of D.C.L. In writing about it to Sedgwick, he remarked
that " the scarlet gown, with which your kind wishes intended to clothe me at Cambridge, has come at last among
the Oxford dons. Science is, I am sorry to say, still very
much depressed there. In fact old Gaisford's saying has
come to pass, ' Buckland is gone to Italy, and we shall
hear no more, thank God, of this Geology !' "

The British Association assembled this year at Belfast.
" The meeting," Murchison writes, " was really successful,
and in every way good, except that I have been too much
travaillé by dinners, speeches, etc., particularly in beginning
with an awful feast to the Lord-Lieutenant, at which the
Mayor presided—twenty-two toasts, and a sederunt from
six till one !"

After the meeting he took a tour in Ireland with his
friend De Verneuil and a Russian traveller, whom he had

brought to the Association. Of that Russian acquaintance
(who, it is needless to add, was not his esteemed friend
Von Keyserling), he records the following anecdote :—
" When we went to a great evening party to meet the
Viceroy, Lord Eglinton, at Sir H. Bateson's, my friend
T— fell much in love with the pretty wife of the General
commanding, and thought he had made a conquest, inasmuch
as she asked him to dinner next day. There he went, but
what was his woe when all the lords and ladies having gone
out from the room, he was left to come out with the parson !
He came to the great ball afterwards, where I found him
sulky as a bear. It was in vain to explain to him that all
the people there except himself had high titles. He thought,
as a stranger, he ought to have been preferred to all. I had
afterwards a good opportunity of gratifying his pride. When
we were visitors to Lady Londonderry, at Garron Tower, on
the coast of Antrim, I happened to meet Lady L. first, when
she said, ' But where is the Russian Prince ?' I replied, ' I
am charmed to hear you give my friend that title, and if you
will only treat him as a Prince, all will be set to rights.' I
then of course explained how he had been offended. Now
it so happened that the General and his wife arrived soon
after, and also many notables. But when dinner was announced, Lady L. gave her arm to T— and led off. Never
was a man so enchanted, and in the evening he said to me,
' You see, my dear fellow, when one comes into really good
society, one's right position is at once recognised !' "

The brief tour would appear to have been meant more
for the purpose of showing De Verneuil and the Russian
" Prince" a little of Irish life and hospitality than for
geological ends. But of course rocks, soils, and bogs were

looked at on the way.   Lady Murchison had not crossed to
Ireland, but awaited the return of her husband at the Menai
Straits, where, with his companions, he rejoined her.   A
week was spent in leisurely crossing Wales by way of some
of the more interesting geological sections.   The sight of the
Welsh mountains, and partly the furious storms of the
equinox, frightened their Russian friend, who had again lost
his heart at Dublin, and was now in no mood for prosaic
geology.   So we find the following entry in the diary :—
" Took leave of the great T. at the pont of Aberglaslyn, after
he had gone round Snowdon in a car, and in his Parisian
boots.   The good De Verneuil stuck to me, and, in advancing with him, we soon got to work.   Admirable proofs
of grinding and rubbing action on the rock bosses which
advance into the flat plains of the ancient bay of Tremadoc."
They visited Mr. Ramsay, and were taken by him to the
flanks of Moel Wyn to see the Lingula flags ; ascended
Cader Idris in a bright morning after the storms, with the
Geological Survey map in their hands, turned thence to the
hills of Old Radnor, and so down by Hereford to the sections of the Old Red Sandstone on the Wye.   They reached
Belgrave Square on the last day of September.

But that the return to town did not imply an immediate
return to geological work is shown by a note written next
day to Whewell :—" I have just arrived from Ireland *viâ*
Wales, having (to the great satisfaction of my wife) been
obliged to travel 150 miles from the Menai Straits in our
own carriage—a strange thing this to say in this railroad
age.   We came through Radnor, Hereford, and all my old
ground—De Verneuil with us.   I will send you a last word
on the Silurian rocks.   At Belfast there was but one opinion

clearly expressed by every geologist present, viz., De la Beche, Griffith, Portlock, Nicol, and many others. . . . My wife is in the country, and I am going to join her, to shoot pheasants."

To complete the details needful for the completion of the new Silurian volume, further work in the field was found to be called for. Especially necessary had it become to re-examine some of the original sections in the Silurian country, regarding which Sedgwick on the one hand, and the officers of the Geological Survey on the other, had come to views different from those which appeared in *The Silurian System.* Accordingly, part of the year 1853 was given to this task. After a dutiful visit to the old General, his uncle, at Cheltenham, and the customary exchange of Peninsular reminiscences, Murchison repaired once more to the Glouces-tershire hills. " I went," he says, " to look over some of my old ground, partly with Hugh Strickland, viz., May Hill, and the cuttings of the new railroad around the south end of the Longhope ridges. I then cautioned him of the danger of working on those lines. It was prophetic of his fatal end, when he was cut in two pieces at the mouth of a tunnel.[1] I also made some detailed sections near Ludlow, railroad cuttings there having laid open new junctions—pottering work, but necessary in these pottering times, the chief work being over."

[1] This melancholy event happened only a few weeks after the ramble mentioned in the text. After the Hull meeting of the British Associa-tion, this amiable and accomplished naturalist was engaged in examining a geological section at the mouth of the Clanborough tunnel on the Great Northern Railway, when he met with his sad and sudden death. He heard a locomotive's whistle, and looking along the line saw a train com-ing up, on which he stepped across to the other set of rails ; but at the same moment an express train dashed out of the tunnel upon him.

A good deal in the way of the comparison of geological
sections of the palæozoic rocks, especially those of Permian
age, remained also to be done on the Continent before a
broad general picture of the succession and diversities of
these rocks throughout Europe could be given in the chapters
of the forthcoming work.   To get this work done another
tour in Germany was undertaken.   Mr. (now Professor) John
Morris accompanied Murchison on this expedition, his prac-
tised eye and wide knowledge of fossils being likely to give
great assistance in the comparisons about to be instituted.

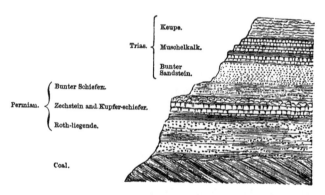

PERMIAN ROCKS OF GERMANY.

The route lay up the Rhine into Westphalia, and through
Cassel to Leipzig; then into the Thüringerwald, and back
by Freiberg into Bohemia, to see Barrande and his geology
at Prague.   Thence a rapid journey northwards brought the
travellers to Berlin, where they met Humboldt, G. Rose, and
other kindred spirits.   Turning southward and westward
again, they hammered through the Devonian rocks, taking
Frankfort by the way, as far as the western end of the
Taunus.   Reaching thus the Rhine again, they looked once
more at its rocks, and descended its course until they struck

through Belgium to Paris, and so home. "In all fifty-three days absent," Murchison remarks, "and how much I have seen!" The more important results of this busy tour subsequently appeared in a joint memoir by the fellow-travellers read before the Geological Society.[1]

Among the geological details of the note-book some gossip about the illustrious Barrande occurs. Thus at Carlsbad the following entry is made :—" The Silurian hero of Bohemia and Germany, my dear Barrande, is fortunately here, and has enabled me to add to and correct my note-book of 1847. The walks and talks with him are most effective. What a noble character! Then we have here Lovén the naturalist, from Stockholm, with his mild, gentle manners, so that Morris is as delighted as myself. Life very primitive. Up at five, to bed at nine. Harvest beginning, and people leaving fast ; most of the great folks gone. . . .

" *Prague.*—To give some faint notion of the money paid to workmen employed by M. Barrande, and furnished by him with lenses and other instruments to detect the minute forms of those metamorphoses [of Trilobites], the two drawers containing the specimens from which the figures have been taken have cost the author 5000 francs; the two genera *Arethusina* having cost not less than 10,000 francs. . . .

" Having ransacked M. Barrande's brains and his noble collection, we also had from him and his old *factotum* woman a capital dinner in his little apartment ; there being just room for the table, among his chests of specimens which are monstrously piled up in great masses, but all in *lucidus ordo.* Although he literally knows where every specimen lies, he has enough of them for a hundred inferior collections,

---

1 *Quart. Jour. Geol. Soc.,* xi. (1855).

but can only make one superior and unique, which I hope to obtain for the British Museum." [1]

Shortly after reaching England Murchison sent a long narrative of this ramble to Sedgwick. Its voluminous detail, which even Sedgwick confessed himself for want of good maps unable to follow, would be out of place here. But a few passages may be quoted showing, as they do in a touching way, the writer's tender regard for his friend, and his unwillingness to believe that any lasting estrangement could ever arise between them.

" *Oct.* 13, 1853.—MY DEAR SEDGWICK,—As this is about the time you betake yourself to the College walls, I now write to tell you of some results of my late journey in Germany, being sure that they will interest you deeply. In fact I was going to write to you from Berlin, where I went to visit old Humboldt, and to tell you of the *dénouement* of our old region of Saalfeld and adjacent tracts of Thüringia and Saxony. I intended that the *annonce* should have reached you in your geological chair at Hull,[2] if only to show to the world that no bickerings about Lower Silurian and Cambrian interfered with our friendship. But you know what foreign travel is, and how ever moving and fidgety a fellow I am, and really I had no time. Besides, I had then only passed by Bonn, Cologne, and the grand new railroads which have so finely laid open all our old sections along the Westphalian frontier. If ever you go there again, and take the *Eisenbahn* from Düsseldorf, you will stop at every other station, and between them be whisked through the grandest slashes possible which have been made in our Devonians,

---

[1] This was subsequently effected.

[2] Sedgwick filled the Chair of Section C at the British Association meeting this year at Hull.

etc. I had not then revisited Frankfort, Wiesbaden, and the Rhine, and had not conversed with the Sandbergers and Von Dechen and others. So I felt that I should be premature, and perhaps erroneous. Now I have it all in hand, and have written a demi-chapter and tabulated it all for my book.. . . .

"I was some days at Prague ⸱with Barrande, besides the twelve days we spent together at Carlsbad, where most fortunately I found him. His collections are more marvellous than ever. But the great point of interest for you, as for myself, is to know the final result of all the Rhenish business. As we went thither like Luther and Melanchthon (I hope not like Calvin), to reform the old 'greywacke creed,' it is right that we should show to what extent we also erred. Not that any of our sections were erroneous— not that the chief physical masses are not as we placed them; but simply that we were wrong in applying 'Silurian' to that which has proved to be Lower Devonian. On that point I take to myself naturally the greatest blame. . . .

"I have given up an hour or two, though very very busy in condensing all my 'foreign affairs,' to have this chat with you on things upon which we must have a community of feeling and thought. In looking over our old publications and my old memoranda-books, we seem to be such complete Siamese twins that it does my heart good to turn to them and pass away from all the *irritamenta* about a nomenclature which has led too many persons to think that we were estranged. I will never go on wrangling. What I have done, and said, and published, has never impugned the accuracy of your labours in the field, and I only regret that some expressions that have fallen from you about my mistakes and errors should have appeared harsh to others. Why, there is no geologist alive or dead who has not made plenty of mistakes, and though I have never alluded to those of your omission or commission, other persons have done so.   At all events, whatever be the nomenclature adopted, we mean the same thing : our views on the progressive creations, on the true order, on the infinitely greater intensity of former causation—on these and numerous other points we are agreed, and my object in writing to you is to show how well our Devonian views have been eventually worked out upon the Continent. . . .

"Morris was an excellent companion, and of great use in the cabinet.   I have also made the Permian stronger, and there again I revert with true pleasure to your very masterly memoir on the Magnesian Limestone, in which you gave the true order of the natural group.   Some of the Germans wish to keep that cursed Rothe-todte out of the group, but I will not hear of it, and Naumann, Geinitz and others have already termed the whole 'Permische.'   I shall be here [Tunbridge Wells] as head-quarters till Christmas probably, or merely going up and down to meetings only.   My wife sends her kind love, and hopes to hear nothing of your ailments, and that you are quite well.—Believe me to be, my dear Sedgwick, your fast friend,   ROD. I. MURCHISON."

To this long letter, Sedgwick sent a long answer.   The generous spirit in which he could still meet the advances of his friend may be judged from a single extract :—" I hope you won't think my last letter ill-tempered.   If so, set it down to the fiend Gout.   I am delighted with the tone of your letter. It is frank and friendly, as it ought to be, and as your letters used to be. Two or three things helped to set my back up. I know that I am a great procrastinator, partly from temperament, partly from multitudinous engagements that pull hard at me, and chiefly from a condition of health which for months and months together makes writing and sedentary work very very irksome, and almost impossible. Still, though a man is behind time with his rent, he rather grumbles when he finds on coming back to his premises that a neighbour has turned out his furniture, taken possession, and locked the door upon him. This is exactly what you did." And then he reopens the question in the wholly groundless belief that his friend had " stolen a march upon him," and in total forgetfulness of the fact, which has already been proved in this narrative, that Murchison actually consulted him and made him privy, by sending both MS. and proof-sheets, to all that he was doing and meant to do. Sedgwick's confession as to his own dilatoriness is valuable, for that temperament of his led directly to the whole of the dispeace.

In the October of this year died, at Bath, General Sir Alexander Mackenzie—the military hero of Murchison's boyhood, the guardian and commanding officer under whom he passed his youth, the friend and companion of his riper years. Death had recently been busy severing the links that bound the matured geologist with the old Peninsular days. No tie had connected him more closely and fondly with them than that which was now broken by the decease of his uncle.

Before passing from the record of this year we may linger for a moment over the memory of another friend

whom it carried away into the past—Leopold von Buch.
This illustrious man has already appeared in the pages
of this biography, but mainly under the aspect of oddity
and eccentricity, which was the aspect under which he
first appeared to casual observers. But it is a pleasure
as well as a duty, in quitting the narrative of the busy
years during which his active spirit traversed Europe, to
bear a tribute, however humble, to the genius wherewith
he lighted up every branch of science which he touched.
His mental vision was as wide as it was definite and
clear.   Admirably conversant with details, he was yet
gifted with that far-searching philosophic spirit which ever
strives to look through the scattered facts up to the laws
which govern them.   Probably no geologist of his time had
so wide a range of knowledge and acquirement.   He was
great and original in physical geography, in dynamical and
stratigraphical geology, in palæontology.   In each one of
these branches of science he was a pioneer, seeing far into
the darkness, and casting in front of him the clear light of
his own genius to guide the way of subsequent explorers.
Personally, too, with all his idiosyncrasy, impulsiveness,
and quickness of temper, he was at bottom one of the
largest-hearted of men, full of tenderness and generosity, care-
less of himself, sympathetic, and actively benevolent towards
others.   With no family or official ties, and possessed of
an income which, though slender, sufficed for his moderate
wants, he had rambled all over Europe, everywhere making
the acquaintance of those who followed the same pursuits,
and evincing a personal interest in them.   And thus when,
at the ripe age of seventy-nine, Von Buch, active and indomitable to the last, was gathered to his rest, there was

hardly a corner of the globe into which geology had made its way where his death was not felt as a personal loss.

By the decease of General Mackenzie Murchison received a very considerable addition to his fortune. In the meantime, however, the event involved him in much extraneous work which sadly interrupted the progress of his literary labours. Writing to his friend Mr. Murray he says, "The decease of my uncle, with the business it has necessarily thrown on me, to say nothing of the Bellot Testimonial, with which they saddled a willing horse, have shut up *the book* for ten days. Moreover, the printers have sent me nothing, whilst people tell me that you have advertised it as coming out! Five entire chapters are not written, and the index is not begun, so that the issue of the work, in its entirety, this autumn, is utterly impracticable."

The summer of the following year (1854) had begun before the book was actually published. At last it made its appearance as a stout octavo volume of 523 pages, with abundant woodcuts and plates of fossils, under the appropriate and euphonious title of *Siluria*. It was dedicated to De la Beche, by whose labours and those of his associates in the Geological Survey the area of the Silurian kingdom has been so effectively extended.

As the publication of this work marks another stage in the progress of British palæozoic geology, it may be desirable, in conformity with the plan followed in the foregoing narrative, to take here a rapid and general view of what had been done in that department of science since the appearance of the original *Silurian System*. The number of labourers had so increased, and the field of their work had now been so widened, that any adequate review of this subject must necessarily lie wholly beyond the scope of a biography. Still a mere outline will be useful in enabling us to see where in his progress Murchison now stood, and what his relations were to the onward march of his favourite science. The non-geological reader, to whom a *résumé* of the stages of advance, already in great part noted in previous pages, is irksome, may pass over the remainder of this chapter. To the geologist, however, it may not be without advantage to pause here for a little to cast a glance backward, with the view of distinctly realizing the point at which palæozoic geology had now arrived.

At the time when the *Silurian System* appeared, the rocks which had once been classed under the vague terms "transition" and "greywacke" were grouped in England into two great series. Of these one had been elaborately worked out by Murchison, and had received from him the name of Silurian. The other, stretching through the mountainous regions of Wales and Cumberland, had been resolutely grappled with by Sedgwick, who, after partly unravelling the intricate structure of North Wales, termed this series Cambrian. It was believed, as we have already seen, both by these two observers and, on their authority, by the rest of geologists, that the so-called Cambrian rocks lay deep beneath the various groups into which the Silurian masses had been divided. No distinctive fossils had been found in them, and in this respect, admirably as their physical structure and mineral subdivision were worked out by Sedgwick, they failed to offer the same facility for comparison with other regions which the well-characterized suites of fossils gave to the Silurian series. Fossils had indeed been collected by him from his Cambrian masses, but they lay packed away for years, and when at last examined they were found to present a wholly Silurian character. Still it was hoped that more extended research might yield peculiar groups of fossils, and thus enable zoological as well as lithological lines to be drawn through the vast mass of Cambrian strata. This hope had not been realized.[1]

In course of time, however, as detailed in the foregoing chapters, various other observers began to look at the arbitrary line of division which had been drawn between the two territories of Cambria and Siluria. That line, drawn in part by Sedgwick himself, was not based on any natural feature or series of sections.[2] It was inserted chiefly as a conventional boundary to separate the tract which Murchison knew and had named Silurian from that which he had not examined, and which he supposed to be occupied by the older or Cambrian group of rocks. When it came to be scrutinized on the ground it was found to be wholly illusory. Mr. Bowman (1840), and afterwards Mr. Sharpe, Sir Henry de la Beche, and Mr. A. C. Ramsay[3] (1842), had shown that the Silurian mineral characters and the Silurian fossils ranged far beyond the supposed line of demarcation into the so-called Cambrian region, and therefore that at least to some extent the Cambrian rocks were undistinguishable from what had elsewhere been termed Lower Silurian. Availing himself of these sugges-

[1] See *ante*, vol. i. p. 382, *note*.

[2] That for the boundary-line Murchison was not wholly responsible, as Sedgwick maintained, is shown by the letter already quoted (vol. i. p. 307), where Murchison, in writing to Sedgwick, actually alludes to the boundary-line in North Wales having been inserted by the latter, and evidently speaks of a fact which at the time was fresh in their recollection.

[3] It was in the month of June 1842 that Mr. Ramsay found fossils at Llandeilo, proving this extension, and the same fact was made out one day later, but independently, by Sir H. de la Beche, at Llangadoc.

tions, and strengthened by the evidence which extensive
foreign travel had brought before him, Murchison (1841 and
1842) declared his belief that the conventional line set up
between his territory and that of his friend Sedgwick had
no longer any geological significance, and that the term
Cambrian must cease to be used in zoological classification,
seeing that the fossils found in the so-called Cambrian rocks
were proved to be of Lower Silurian types.   Subsequently
(1842-6), the Geological Survey, in the course of its extension into Wales, ascertained that the very same strata which
had been termed Lower Silurian by Murchison extended
throughout South Wales, and they were afterwards equally
recognised all through North Wales.   De la Beche and his
colleagues thus proved that the terms Cambrian and Lower
Silurian were in a great measure two names for the same
series of rocks.

Geologists had then to determine which of the names
should be retained. Of the two terms, Silurian had been
based not merely on mineral characters, but mainly upon
fossil evidence.   It had therefore been capable of adaptation
in other countries, and indeed, as we have seen, had been
actually applied to the rocks throughout many widely
separated regions in the Old World and in the New.   The
term Cambrian, on the other hand, as *originally* used,
described a vast succession of strata divided into groups, not
at first by fossils, but by mineral characters—the only kind
of classification possible in the absence of fossil remains,
but one which is apt to possess a merely local significance,
and to be therefore incapable of general application.   This
is a fact which must be kept clearly in view.   Sedgwick,
when he gave the name " Cambrian " to his Welsh rocks,

SIR HENRY DE LA BECHE, F.R.S.
From the Engraving of the Enamel by H. P. Bone.

though he knew they contained some fossils identical with known Silurian species, had only mineral characters on which to base his subdivisions, and to offer for comparison with the rocks of other countries. But the Silurian group- ing was founded on fossil evidence, that is, on the general history of life on the globe. It could be, therefore, and it had been, successfully applied to the rocks of widely sepa- rated countries.

When the Silurian classification had been generally ac- cepted and extended into foreign countries, Sedgwick, after a long interval, returned (1842) to the study of his so-called Cambrian rocks. Admirably did he unfold their physical succession step by step through the rugged region of North Wales. He had accepted without sufficient examination the assumption that Murchison's groups were all younger than his own, and finding it difficult to reconcile this postulate with the facts, had in vain tried to make his sections fit satisfac- torily into those of his friend. When at last it was discovered that the assumption had been an error, he attributed it to Murchison, declaring that the latter had misplaced his groups, and claiming therefore the Lower Silurian rocks as properly part of the Cambrian domain. But even had the author of the *Silurian System* been wholly answerable for the mistake, this could not have affected the indisputable fact that the order of Murchison's formations in the original and typical Silurian region remained, and still remains, as he placed it. Sedgwick's vehement contention about the misinterpretation of the relation of the Llandeilo flags to his Upper Cambrian rocks, had really hardly any bearing on the general question of nomenclature. The Silurian arrange- ment was right; the application of it to North Wales had

at first been wrong, but this error could not make wrong
what had been shown to be right, nor could it affect Murchison's priority in having established a true palæontolo-
gical classification.   He maintained that, as he had already
made out the Silurian classification, rocks which proved
to be the same as his Silurian groups could not possibly
receive any other name.   By general consent, the verdict of
geologists all over the world has been in Murchison's favour.

It is impossible, however, not to feel, and in such a
narrative as the present not to express, a true and deep
sympathy with Sedgwick.   He had given the labour of
some of the best years of his long and honoured life to the
disentangling of the structure of his favourite Cambrian
rocks.   He had far more serious difficulties to grapple with
than his friend ; these, by dint of patience, courage, and his
own genius for physical geology, he had successfully over-
come.   It was hard, therefore, to find that after all his
labour in conquering it, the stubborn territory was claimed
by another, who had borne no part in its subjugation.
Such, however, was the fortune of war.   Murchison, by the
laws of fossil evidence, had established his claim to all
territory peopled by his Silurian types of life, and when
these types, and these only, were afterwards found in Sedgwick's domain, that domain fell naturally and inevitably
into the empire of Siluria.

Passing from the general question of boundary to the
internal development of the Silurian domains since 1839,
we find that considerable progress had been made before
1854 in two departments, viz., in the order and grouping of
the rocks, and in the enumeration and description of the
fossils.   In this progress Murchison himself had scarcely

any share, while, on the other hand, Sedgwick played a foremost part. The latter geologist not only did admirable work in the field himself, but by employing as his assistants such men as M'Coy and Salter, who brought trained eyes to the identification and discrimination of fossils, he rendered the most essential service to this branch of British geology. The general order of succession determined by him among the older rocks of North Wales remained unchanged, though his term "Cambrian" was now very generally reserved for the massive, and up to that time unfossiliferous, strata lying below the fossiliferous Lower Silurian rocks. Taking advantage of mineral characters, he had arranged the rocks into great groups. He showed that overlying the Bangor group, to which other geologists were disposed to restrict the term "Cambrian," lay a well-marked zone full of *Lingulæ* and Fucoids, to which he gave (1846) the now well known and constantly used term of "Lingula Flags." At the same time he made out that over these flagstones lay another distinct zone, marked out both by mineral characters and fossils, and to which he gave the now familiar name of "Tremadoc slate." Both these zones lay beneath Murchison's lowest Llandeilo rocks, from which, though still showing, as some thought, a Silurian facies, they were distinguished by a general want of community of fossil species. Thus the establishment and naming of the lowest zones of life up to that time detected in Britain was the work of Sedgwick.[1]

The Llandeilo group, the basement zone of Murchison's series, had now been traced by the Geological Survey over

[1] The Lingula flags of the Longmynd country had been mapped by Murchison with his Llandeilo group. He had hitherto found no fossils in them, and of their extension as a great group into North Wales, he was of course ignorant.

a great part of Wales. In like manner the Caradoc group
had been found to spread over a wide district, and to be the
same as part of what Sedgwick had termed the "Bala
group" of his Upper Cambrian. But in the course of his
work in the field that geologist had found (1851) by the
evidence of fossils, that Murchison, and after him the Geological Survey, had probably grouped more than one distinct
geological zone under the common term "Caradoc." In consequence of this suggestion, the Survey, on re-examining
the ground, found Sedgwick to be right. The upper portion
of the Caradoc sandstones turned out to lie unconformably
upon the rest in Shropshire, and to show quite an Upper
Silurian character in its fossils. This upper part was at
first known as the "Pentamerus Beds." Murchison, as long
as he could, resisted the splitting up of his original subdivisions. He had himself failed to detect any break in the
Silurian succession, and now that such a break was proved
both by physical and organic evidence, he strove to show
that it was after all only a local phenomenon. Yet it was
certainly general all through the Shropshire and South
Welsh regions. The true history of the Pentamerus Beds
had, however, not yet been ascertained.

With the exception of a shifting of the place of one of
their characteristic limestone bands, the Upper Silurian
groups remained as they appeared in the *Silurian System*.

While all this labour had been bestowed upon the elucidation of the order of the rocks, another band of workers
continued busy with the fossils, which had been obtained in
far greater numbers than at the time when the *Silurian
System* appeared. In addition to Lonsdale, Sowerby,
Phillips, and the others who had taken part in the determi-

THOMAS DAVIDSON, F.R.S.

nation of the original Silurian species, younger men of equal ardour and industry had arisen in this country. Foremost of these stood Edward Forbes, Morris, M'Coy, Salter, and Davidson. By their researches the number of species in the older rocks of Britain had been largely augmented, and new and ever increasing light had been thrown upon the history of the earlier forms of life.[1]

Another great addition to geological knowledge since the publication of the *Silurian System* was the establishment of the Devonian System, as already narrated. That addition had been obtained by the co-operation of Sedgwick and Murchison, with Lonsdale and Sowerby. The separation of the Permian System from the general mass of the lower mesozoic red sandstones, and its insertion as the uppermost member of the palæozoic systems, was likewise a notable change.

Our brief retrospect, however, would be very incomplete if it took no account of the remarkable way in which the palæozoic classification, established in Britain, had been extended into other countries. The share which Murchison and Sedgwick had in this extension has been traced in the foregoing chapters. It would be out of place to attempt even a bare enumeration of the names of foreign contributors; but those of De Verneuil and Barrande in Europe,

[1] Among the more important contributions to the palæontology of the older rocks of Britain, which had appeared since *The Silurian System*, may be noticed Phillips' Memoir on *The Malvern Hills;* the *Decades of the Geological Survey*, containing the descriptions, by Forbes, of the Echinoderms, and, by Salter, of the Trilobites of the Silurian rocks; the Fasciouli of the *British Palæozoic Fossils*, by Sedgwick and M'Coy ; and the descriptions of Silurian Brachiopoda, by Davidson, in the *Bulletin* of the French Geological Society, and in the *Monographs of the Palæontographical Society*. Besides these large works, there appeared numerous papers in the Quarterly Journal of the Geological Society, and in other scientific periodicals.

and of Hall and Dale Owen in America, cannot be omitted.
In all parts of the world the Silurian type of life had been
recognised as that characterizing the oldest fossiliferous
rocks. A vast number of new species had been described as
occurring in the rocks of other countries. Nevertheless, the
distinctive general character remained which Murchison had
recognised as pervading the rocks classed by him under
the term " Silurian."

Such then was the general aspect of this subject when
he published his *Siluria.* That work gave a fair statement
of the state of knowledge at its date, and combining, as it
did, the substance of so many scattered memoirs, it proved of
great service in promoting the methodical study of the older
rocks. Three editions have appeared, each removing some
of the imperfections and errors of the first, and proving, by
their steady sale, the useful part which the work has taken
in the geological literature of the time.[1]

It was with the view of embodying all this progress in
one comprehensive narrative, that the author of *Siluria*

[1] It would be far beyond the scope of such a work as the present biography to enter into much detail in matters of scientific controversy.
Enough (some readers may be disposed to think, more than enough) has
been already said on the subject of the Cambrian-Silurian dispute. It
may be mentioned, however, in this foot-note that the appearance of
Siluria revived the bitterness in Sedgwick's mind. He believed himself to have been most unjustly treated by his old friend, and he gave expression to his feelings in language of a vehemence seldom seen in scientific writings. (See *Phil. Mag.* for November 1854, and *Introduction to
British Palæozoic Fossils.*) He called on Murchison to express his regret
for having fastened an unmerited accusation on him, and in the strongest
terms denounced what he considered to be the geological mistakes of the
author of the *Silurian System,*—mistakes now reproduced and aggravated
in *Siluria.* Most men would have resented such language and refused to
hold further intercourse with the man who could deliberately print it.
Murchison could not apologize for an act of injustice which he felt he had
never committed, but he continued to use every means of pacifying his
friend, of whom he still spoke and wrote in the same old affectionate
terms.

prepared the chapters of that work. Evidently one great contrasting feature between the original *Silurian System* and this successor to it lay in this, that the one work was based mainly on the author's own original labours, of which it was the fresh and detailed expression, while the other consisted partly of a re-statement of these early labours, partly of an account of their extension by the author himself, in his own country and abroad, and partly of a *résumé* of what had been contributed to the common fund of knowledge by other fellow-workers. Valuable, therefore, as *Siluria* was, and eminently useful as a compendium and indispensable *vade mecum* for students of the older rocks all over the world, it lacked the freshness and originality of the earlier work. Nor was it so easy to distinguish between what had been achieved by the personal exertions of the author himself and what had been worked out for him and with him by other fellow-labourers pressed into the service. Not that he withheld acknowledgment of assistance; he frankly admitted his indebtedness, and entertained very grateful feelings towards those who helped him. But general acknowledgments furnish little clue to the appraising of the relative value of each workman's share in the building up of the edifice. Fortunately for the cause of progress, this personal element is of but small moment. The temple of science is ever growing in height and breadth, and though, like the mediæval masons, each builder in that temple may wish to leave his distinctive mark upon the stone which he has conscientiously and lovingly laid with his own hands, he must in most cases be content with the purer satisfaction of seeing the rise of the great building to which he knows that he has added something.

CHAPTER XXII.

HAVING launched his volume, the author of *Siluria* had proposed another Continental tour to supplement that of the previous year, and enable him to improve, correct, and extend the account of the German geology in that work, which he knew to be capable of much amendment. But the execution of this design was first retarded, and then, though carried out, much curtailed by a succession of calamities foreign and domestic. After an ominous darkening of the political horizon in Europe, war had at last broken out between Russia and Turkey, into which Britain and France soon threw themselves. That our country should declare war against the Emperor whom Murchison so passionately admired filled him with surprise and sorrow. He made no secret then among his friends, though he found himself in a meagre minority, that he thought the war foolish and unnatural, and so he continued to think as long as he lived. Writing to M. Barrande immediately after the declaration of war, he says,—" I cannot tell you how much I have been grieved and irritated by these untoward events. I had quite settled it in my mind that Russia and England never could be estranged ; and as most of my countrymen know my feelings, too many of them have often the bad taste to jeer at me on my sore point. But whilst I am as loyal a subject as any one of my own Sovereign, and heartily pray for success to her arms, I as devoutly pray for peace, and shall never cease to regret that we should be at war with an ancient ally of near 300 years' standing. . . .

" I have given my best and arranged Silurian collection to the British Museum, *because* the Trustees have purchased yours. . . .

" I think of leaving London either the 21st or 23d July, my wife being at Lichfield House, Richmond, for the summer. If you get this soon you may write to me, Clausthal, Harz, whither I go to see Ad. Roemer, and ascertain if he really has any true Silurian there. I shall also endeavour to fortify my Permian foundations by re-examining the *Todte-liegende* of various tracts. This operation will probably take me to the south of Breslau."

But the excitement of war-time, added to his own chagrin at the breaking out of such a contest, was not the only cause which kept his movements uncertain. His brother, to whom he was bound by a life of tender affection, had been ailing, and the state of health of the invalid was now such as to cause much anxiety. In spite of all their geological bickerings, there was still no one to whom Murchison could so openly unbosom his innermost thoughts, or turn so instinctively for sympathy as Sedgwick. When the prospect looked gloomiest, he thus wrote to his old comrade :—

" Alas ! my dear friend, I am in a grievously afflicted state, and quite unequal to much business. My only brother Kenneth, to whom I am sincerely attached, is stricken with death, and cannot survive many hours, or days. His malady (a heart disease) has made frightfully rapid progress in the last few weeks. I am myself bent down by a vile influenza, which has left me weak, and my doctor has ordered me abroad. But for this unfortunate state of affairs I should have been on the Brocken. I had resolved to see again our old ground, and satisfy myself, *in situ*, as to whether there are any fossil-beds older than the lowest Devonian."

Three days afterwards he writes again :—

" I had no hope, but certainly was unprepared for the rapid dissolution of my poor brother, who died at half-past four on Tuesday, 1st of August, in my presence. He conversed a little with me even a few minutes before, and he only struggled to live on in hope of seeing his only living sister, Mrs. Hull—but in vain. This is a sad shock for my worn-out nerves, which required a cheerful summer tour, and now I am nailed down by all sorts of business, every act of which reminds me of my loss. We have been loving brothers through life, ever since he came, at five years old, to Durham school, where I, being two years older, had to fight his battles."

It was under the melancholy of this bereavement, and with the further depression caused by indifferent health, that Murchison, by medical advice, again set out for the Continent. It could not be a mere tour for gathering renewed energy. So long as a hammer could be carried or a section be visited he must needs turn his excursions to a geological use. Hence he chose this autumn a region where, amid much varied scenery, he could find plenty to interest him among the rocks—the Harz and Thüringerwald. He was again accompanied by "the trusty John Morris of Kensington."

The gorges and scarps of the Harz, bristling with their dark needles of fragrant pine, remained still what they were in the early days when the two knights of Cambria and Siluria climbed their slopes. But of their rocks more could now be said than was possible then. Even into these quiet valleys modern geology had made its way, and resident observers had gathered collections of the fossils. It had become therefore an easier task, with these local aids, to group the rocks and fix their relations to those of other parts of Europe. How Murchison and his companion now did this, and how, continuing and improving their work

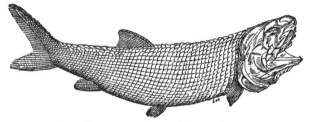

FOSSIL FISH FROM THE PERMIAN ROCKS OF GERMANY
(*Palæoniscus Freislebein*, Ag., Mansfeld).

of the previous year, they extended their labours through the Thüringerwald, they have themselves told in their conjoint memoir to the Geological Society.[1] Abstracting the geological matter which went to form part of that essay, there remains little of any personal interest in Murchison's notes of his journey. It is clear enough, indeed, that before he had been many days at work he had forgotten all about his ailments, and become as thoroughly engrossed as ever in " Spirifer-sandstone," " Stringocephalus limestone," Melaphyre, Permian, Bunter, Trias, and other cognate subjects. That he had likewise regained his bodily energy may be

[1] *Quart. Journ. Geol. Soc.*, vol. xi. p. 409.

shown by a single extract: "Walked for three hours, to the summit of the Brocken. Good weather for ascent; thin floating clouds. The old pillar on the summit is gone. Here I write in my sixty-third year, having walked up the 4000 feet as well as I did in 1828 with Sedgwick."

The travellers returned in time for the British Association meeting at Liverpool. Of that meeting Murchison has left a few memoranda : " What a change for the worse in travelling ! Locked into a first-class carriage, with a roof scarcely above my head and cushions half an inch thick. . . . I now found that the savans were as usual to be made the tail-piece of another great meeting—the opening of the grand Music Hall. . . . Much that followed on other days was mere display and speaking *ad captandum vulgus ; rechauffés* of old meetings, and little new. Men of intellect can employ themselves better than in teaching women how to begin science. . . . I gave, to the Section C my general views (with great table, etc.) of German classification. . . . In my own Section E, I conducted everything to my satisfaction, and kept the rivals, geographers and ethnologists, well together, getting Drs. Conolly and Latham to preside on their own hobby. I steered, I hope, clearly through some difficulties. On Monday I proposed the Duke of Argyll as our next President at Glasgow. . . . At the grand Presidential dinner in the Philharmonic, attended by seven hundred persons, I was put in the chair, because Lord Harrowby had lost his voice. He sat on my right hand while I made the speech. The best and jolliest thing, however, of the whole was the finish of the meeting at St. Helen's, where the natives gave a dinner to three hundred. It was a really good thing, But this killed me ! "

The last phrase of this extract seemed at one time likely to be realized, for immediately after the meeting he got so ill that his wife, then visiting in Berkshire, was sent for. Recovering soon, he moved southwards, and in November went to Up Park. But to be there at that time of the year and not to have a gun in his hands formed no part of his plans. The result might have been anticipated. He returned to London in a very low state of health. But his admirable constitution carried him through this grand climacteric year ; only thenceforward he abjured a habit for which he had long been celebrated—he ceased to be a smoker.

To all lovers of science in Britain the autumn of this year was clouded by one of the saddest calamities which had befallen the progress of natural history in this country for many a day. Edward Forbes, young, bright, full of promise, and already with a world-wide reputation, died after a brief illness. How Murchison felt the loss he thus expressed at the time to his friend at Prague :—

" MY DEAR BARRANDE,—I have been requested by Mrs. Forbes (through the medium of Professor Ramsay) to announce to you the distressing news of the death of her husband, Professor Edward Forbes. This lamentable event took place at a villa near Edinburgh, on Saturday the 18th of November. In common with all my contemporaries, who loved and esteemed E. Forbes as I did, I mourn over this sad catastrophe, and can hardly realize it to my mind. Six weeks ago, and at a concluding fête given to the British Association at St. Helen's, near Liverpool, he sat on my right hand in perfect health and in the highest spirits ; whilst I, almost double his age, was then suffering from fever and pain, which ended in a severe illness, from which

I have been recovering, and which I well-nigh made fatal
by giving way to the solicitation of my friends to drink
champagne, stand up, and make a speech for the advancement of science.   Better that several old 'sabreurs,' like
myself (who have pretty nearly done their work), should
have passed away than that the bright genius and profound
knowledge of Edward Forbes should be extinguished.   He,
poor fellow ! met his death from neglecting a cold, which he
caught in his last autumnal expedition, and in pertinaciously
insisting upon continuing his lectures when really ill.   Perhaps you will do well to write a few lines to Mrs. Edward
Forbes, Wardie, near Edinburgh, as condolences from such
persons as yourself, De Verneuil and De Koninck (to whom
I write), will be soothing laurels to hang over the tomb of
her illustrious husband.

"I have made all the corrections you have pointed out to
me in my *Siluria,* and will also prepare the table of organic
remains you suggest.   Pray double your criticisms.   You
know how many subjects I had on hand, and how much my
time was engrossed by geographical and other public topics
when I put my Silurian chapters together, so pardon the
result.   150 copies out of 1500 only remain to Murray,
and these will probably be sold next month."

In the progress of this narrative we have now arrived at
the beginning of the last well-marked period in Sir Roderick
Murchison's active career.   Ever since he quitted the army,
he had been wholly unfettered in his movements by any
official work beyond what he chose to undertake in connexion with the different scientific bodies which invited
him to conduct their affairs.   He had now, however, ap-

PROFESSOR EDWARD FORBES, F.R.S.
From a Photograph.

proached a time when he once more placed himself in subordination to Government control, and in that position he remained up to the time of his death.

Sir Henry de la Beche, Director-General of the Geological Survey, after a period of gradually increasing debility, died in the spring of 1855, to the deep regret of all who knew him, and bearing with him the respect of all that wide circle of geologists in this country and abroad who could appreciate the solid and lasting services he had rendered to the cause of science. Before we pass on to the appointment of his successor, we may pause for a few moments to look at the organization of the Survey, and the nature of the work which it was carrying on.

As far back as 1832, De la Beche had offered to supply data to the Board of Ordnance for colouring geologically the maps of Devon, Cornwall, and Somerset. A sum of £300 in aid of this service was charged against the Ordnance Survey of Great Britain in that year, but he himself contributed the remainder and greater portion of the expense. From this modest beginning he gradually gained ground, and at last succeeded in getting his operations recognised as part of the work of the Ordnance Survey. His staff of surveyors formed what was called the Ordnance Geological Survey, of which he became director. At first their work consisted merely in placing upon the maps the relative areas of the various rocks. As early, however, as 1835 De la Beche conceived the idea that the operations of his Survey might become the nucleus round which a really national school of geological and mineralogical science might grow, like the Écoles des Mines of other countries. He was, however, too sagacious a man to go before a British Minister of

State with so ambitious a scheme. He began in July 1835
by representing to the Chancellor of the Exchequer (Mr.
Spring Rice)[1] the desirability of gathering together, in some
public place, specimens of all the economically valuable
mineral substances met with in the course of the Survey,
such as materials for making roads, buildings, or public
works, useful metals, and, in short, all minerals having any
industrial importance. Such collections, he suggested, would
be of great service in embodying a large amount of information of great practical value, and not otherwise attainable.
They should be arranged, he said, with every reference to
instruction, and should be placed under the management of
the Office of Works. These plans received a favourable
hearing from the Government of the time. They were further carried out and improved under successive Ministers,
from each of whom, and specially from Sir Robert Peel,
came assistance and encouragement.[2]

The first home for the incipient Museum was found in
a house belonging to the Crown in Craig's Court, Charing
Cross. No sooner had the place been obtained than presents of specimens from the Cornish friends of Sir Henry
and the Survey came pouring in in abundance. In due

[1] See *ante*, vol. i. p. 228.

[2] Sir Robert Peel took a special interest in the formation and growth of
the establishment under De la Beche. He himself drew up the Treasury
Minute dated 27th December 1844, transferring the Geological Survey from
the Board of Ordnance to the Office of Works, Woods, etc. In that minute
also the Geological Survey of Ireland, which had been begun by Captain
Portlock as part of the work of the Ordnance Department, and afterwards
discontinued, was re-established as a branch of the Survey under De la
Beche. The Chief Commissioner at that time was Lord Lincoln (afterwards Duke of Newcastle), who likewise gave much time and thought to
the fostering of the museum and surveys. In particular, he drew up a
full letter of instructions to De la Beche for the conduct of the various
services united under the guidance of the latter.

time, to make the collections available, and to be alike useful
to the public and the Government departments, there was
added to the establishment in 1839 a laboratory, under the
charge of Mr. Richard Phillips, an able chemist, and a man
whose incomparable wit and humour are remembered with
delight by his surviving contemporaries. The rooms soon
became choked up with contributions, and an adjoining
house was secured. Ere long this additional accommodation
proved insufficient, and some of the collections were stowed
away in Whitehall Yard. At last, when in 1845 the Geological Survey was transferred from the Ordnance Department to the Office of Woods and Forests, the Museum of
Economic Geology (so the aspiring establishment was now
called), then under the supervision of the Commissioners of
Public Works, had, through the liberality of the public
and the judicious management of its guardians, so grown,
and the need for far more ample space became so pressing, that the present building in Jermyn Street was autho-
rized.[1]

---

[1] The following memorandum, addressed by the late amiable and accomplished Lord Carlisle, who at the time was Chief Commissioner of
Woods and Forests, to Sir Charles Wood (now Lord Halifax), then Chancellor of the Exchequer, with the appended reply of the latter, has come
into my hands, and may be given here as having some interest in connexion with the founding of the Jermyn Street establishment :—
"It has been settled that my department is to build a Museum of
Economic Geology. The rent of the building in Piccadilly, which the
public will have to pay to the Crown, is calculated at about £2000 a year.
There has been a plan for having shops in the lower storey, which might
bring in £750 a year; but the conductor of the museum, Sir H. de la
Beche, and the economic Radicals, are vehement against said shops; and
Hume says, if I do not assure him that they are given up, he will this
session move an address to the Crown. Will you let me throw over the
shops?"
"If you wish me to denationalize to such an extent this *shopkeeping*
nation, I cannot resist you.   C. W."

The effective force of the Survey, at that time amounting
to only six assistant geologists, was now somewhat increased,
the staff was re-organized, and a branch was established
for the geological investigation of Ireland.   To aid the
Director-General in the conduct of this augmented service two Directors were appointed, Mr. A. C. Ramsay being
selected for Great Britain, and Captain James, R.E., for
Ireland.   Under each of these officers a small separate staff
of assistant geologists was placed, whose duties were to trace
geological boundaries on the maps, to prepare sections, to
collect information regarding the geological structure,
minerals, and fossils of the country, and to assist in preparing the maps, sections, and other publications in which
the work of the Survey was issued to the public.

While this augmentation took place in the working
power of the Geological Survey, an extended scheme was
at the same time adopted, whereby the Museum should become a school of applied science.   Besides the collections,
now forming a noble series illustrative of the rocks, minerals,
and fossils of England, and the chemical laboratory, it was
planned that a metallurgical laboratory should likewise be
instituted.   Full courses of lectures were to be organized on
geology, mineralogy, mining, chemistry, metallurgy, natural
history, and physics, so that a thorough training might be
had, not in mining only, but in any branch of applied science
which might be required.   A room set apart, in 1839, for the
collection and preservation of mining plans, and miscellaneous
information regarding mines, collieries, and quarries, had received the name of the Mining Record Office, and would now
acquire an extended usefulness.   Over the whole establishment
the head of the Geological Survey would preside as Director-

PROFESSOR ANDREW C. RAMSAY. LL.D,, F.R.S.
From a Photograph.

General. Hence, in the union of systematic instruction, with abundant opportunities for practical work in museum and laboratories, as well as in the field with the surveyors, it was confidently believed that the British "School of Mines and of Science applied to the Arts" would be found fully as well equipped, and might be as successful, as any older school of the same kind abroad.

The School and Museum thus organized were formally opened by H.R.H. Prince Albert in November 1851. As Director-General of the whole establishment, it was the duty of Sir Henry de la Beche to take the chair at the Council of Professors, to see that the collections were properly looked after and judiciously increased, to receive the reports of the local directors, to examine, compare, and sanction for publication the geological maps, sections, and memoirs prepared by the field-surveyors, and to visit when requisite any district in which difficulties might have presented themselves which called for his supervision. Moreover, when consulted by various Departments of Government, the Director-General was expected to send in reports upon the value of the mineral products of the colonies and dependencies, to recommend officers to carry out geological surveys in the colonies, and to submit annually a report of all his proceedings to the Department under which he was placed.[1]

[1] The first list of Professors was as follows :—President, Sir H. T. de la Beche ; Chemistry, Lyon Playfair ; Natural History, Edward Forbes ; Mechanical Science, Robert Hunt ; Metallurgy, John Percy; Geology, A. C. Ramsay ; Mining and Mineralogy, Warington W. Smyth. By each of these an introductory lecture was given at the opening of the school, Sir Henry de la Beche leading off with a general inaugural discourse. See *Records of the School of Mines*, vol. i. part i.

In later years other names of note were gathered round him by the indefatigable de la Beche,—Hofmann, Stokes, Henfrey, Hooker, Willis, Huxley, all enlisted by him in the public service, either permanently or

To direct these multifarious operations, and especially to preside over such a band of scientific men, and to control their work, was an office of no small delicacy and responsibility. Admirably had Sir Henry de la Beche discharged his duties. It was his far-sighted shrewdness which created the whole establishment, and to his tact and good-humour it owed not a little of its successful progress, and of the hearty *esprit de corps* which animated its staff.

When this able geologist and excellent man died, the growth and enlargement of the services under his direction led to some difficulty as to the choice of a successor. The recent stimulus given by the Great Exhibition of 1851 to the action of Government with regard to scientific instruction, made it desirable, in the opinion of some well-wishers to the cause of the higher and technical education of the country, that more use should be made of the Geological Survey and Museum as a nucleus round which other branches of scientific teaching might be grouped. That definite action on a large scale would soon be taken in this matter was indicated in the speech from the Throne at the beginning of the Session of 1853, when Her Majesty the Queen announced to Parliament : " The advancement of the Fine Arts and of Practical Science will be readily recognised by you as worthy the attention of a great and enlightened

for such time as enabled them to give their assistance in some special in-quiry.

The Geological Survey, under its two Directors, Mr. Ramsay (Great Britain) and Mr. J. B. Jukes (appointed in 1850 for Ireland), included some able and zealous assistants, among whom were the now well-known names of W. T. Aveline, H. W. Bristow, A. R. Selwyn. In its palæontologist, Edward Forbes, and its assistant palæontologist, J. W. Salter, it possessed two members, each foremost in his own branch of science. In its earlier years, Mr. (now Sir) W. E. Logan had worked as a volunteer under De la Beche.

nation. I have directed that a comprehensive scheme shall be laid before you, having in view the promotion of these objects, towards which I invite your aid and co-operation." In pursuance of this announcement there was drawn up by the Board of Trade, and sanctioned by Parliament, a plan for the establishment of a Department of Science and Art under the Board of Trade, with temporary headquarters in Marlborough House. From that centre it was proposed that a great organization should ramify over the whole country, having for its object the encouragement and fostering of education in Science and Art with special reference to industrial progress. While leaving individual and private exertion unimpeded, the plan contemplated the purchase and distribution, at a moderate price, of models, diagrams, and apparatus for teaching, the purchase of examples for museums, the loan of specimens from the central museum in London, and the obtaining of properly trained teachers for such schools of science and art as each locality might think fit to establish for itself. These benefits were to be imparted by a system in which every district and school in the United Kingdom might participate.

There already existed several independent institutions maintained by Government, either wholly or in part, for scientific purposes. These it was now proposed to unite under the new department. Among them were the Geological Survey, Museum of Practical (formerly Economic) Geology, and School of Mines. In the ambitious programme of the department, the school which had been established by Sir Henry de la Beche, for a specific and important branch of science-education, became the " Metropolitan School of Science applied to Mining and the Arts," and was to form a sort of nucleus round which it was intended that a great central establishment should grow, where every department of science should be taught that might be required for thorough practical training in the industrial arts.

The choice of a successor to Sir Henry de la Beche therefore could not fail to be a serious difficulty to such of the authorities who more especially interested themselves in the development of their comprehensive scheme. Their views are well expressed in the following remarkable document by one of the most far-seeing men who at that time contemplated the future industrial progress of England :—

Memorandum by H.R.H. Prince Albert.

"BUCKINGHAM PALACE, *May* 2, 1855.

"It is important that the opportunity afforded by the appointment of a new Director of the Museum of Practical Geology should not be lost for furthering the general scheme for bringing science and art to bear upon the productive industry of the country, as recommended by the Commissioners for the Exhibition of 1851, in connexion with the appropriation of their surplus and as approved of by the Government.

"According to this scheme, museums of science and art were to be formed on the ground purchased by the Commissioners at Kensington, the main utility of which should not consist in their being a mere collection of curious or interesting objects, but as serving for the purposes of National Education in illustrating courses of instructive lectures.

"With this view, as regards Art, the Schools of Design

were remodelled by a 'Department of Practical Art,' established under the Board of Trade, the museum of which has been provisionally deposited in Marlborough House. Valuable additions have been made to this collection by the recent purchase, under the sanction of Government, of portions of the late Mr. Bernal's collection. Valuable articles are also in the possession of the Commissioners, which are temporarily deposited in Kensington Palace. The whole of this will some day be placed in juxtaposition on the ground with the National Gallery, which is destined to illustrate the history of the Fine Arts.

" With respect to Science, a 'Department of Practical Science' has also been formed, under the Board of Trade; and the institution on which an educational organization would be most easily grafted was that of the Geological Survey and the Museum of Practical Geology, an establishment created solely by the untiring energy of the late Sir H. de la Beche, who raised it, under the Board of Woods and Forests,[1] almost as it were under disguised colours, at a time when little interest was felt generally in the subject, this being at the time the only recognition in this country of the claims of science to be directly fostered by the Government. This institution is now transferred directly to the Board of Trade; a School of Mines has been formed as a branch of it, under Professor Smyth,[2] and the Royal College of Chemistry embodied with it under Professor Hofmann. As it often happens that a person who, through great difficulties and by his own exer-

[1] It was transferred to this Board from its original position under the Board of Ordnance, as already narrated.

[2] Mr. Warington Smyth was not Director, but Professor of Mineralogy and Mining.

tions, has succeeded in realizing one idea does not readily merge this in a larger one, so Sir H. de la Beche cannot be said to have extended the usefulness of his department, but has rather counteracted the plans of the Commissioners by confining his attention to simple geology.[1]

"It becomes of the utmost importance that whoever is appointed now should be made thoroughly aware of the views of Government, and accept the office with a clear understanding that he will be called upon to

[1] The tribute which the Prince bears to the indomitable energy and consummate tact of Sir Henry de la Beche is no more than just. His Royal Highness, however, appears to have been misinformed as to the real breadth of the deceased Director-General's views. He was perhaps the most many-sided geologist of his day in England. He had acquirements in physics, chemistry, mineralogy, and petrography quite unusual among his contemporaries. Moreover, he was an excellent linguist, and for years carried on the foreign correspondence of the Geological Society. His sketches of geological and pictorial scenes were singularly effective and artistic, and to his taste and skill is due the style of section-drawing employed by the Geological Survey. So far from confining his attention to simple geology, he had conceived a great plan for establishing in this country a School of Mines at least as fully equipped as any on the Continent. The wide scope of this plan may be gathered from the mere list of eminent men in various branches of science whom he gathered round him (see p. 181, *note*). It is probably true that he offered or gave no help to the scheme for establishing a general school of Applied Science under the Marlborough House officials. He knew that the school he had himself founded supplied a real want in England, and that it could be successfully carried on. He did not know that the time had yet come for the establishment of a great general school, and though twenty years have passed away since his death, the scheme for the equipment of one great Government School of Science in London has only been partially carried out. It was no small part of his tact to be able to gauge exactly what he could accomplish and to bend all his energies towards it. He felt sure that a thoroughly furnished School of Mines was a thing of vast importance in such a mining country as Britain, and that he could successfully launch it. So far therefore from being really obstructive, he was in advance of his time. To have planned and successfully founded the great Jermyn Street School of Mines was of itself work enough for one man, yet it is only one of the many claims which De la Beche has on the grateful recollection of all lovers of science in this country.

work them out. He should further consider himself not in the light of a simple geologist, but as at the head of a Government educational establishment for the diffusion of science generally as applied to productive industry. Besides the collection in Jermyn Street, which has already outgrown its means of accommodation, there are ready to go to the Commissioner's grounds at Kensington a collection of animal and vegetable produce, now temporarily deposited in Kensington Palace, and formed by the Commissioners themselves; and also a most important collection of mechanical models for the illustration of the history of inventions forming, by Professor Woodcroft, under the authority of the Patent Commissioners, and temporarily stored away in the buildings belonging to the Master of the Rolls. A."

We are here only concerned with the negotiations for the filling up of the vacancy in so far as they are connected with the subject of this biography, and in this respect the story can be perhaps best followed in Murchison's own letters :—

" 18*th April* 1855.—MY DEAR PHILLIPS,—I could say nothing to your pregnant note till now; for it was only last evening (after post-time) that a resolve was come to on my part, which, if you had hinted at it a few days ago, I should have viewed you as charging me with folly as regarded my health, happiness, or peace of mind. But the Rubicon, or rather the Teme is passed, as far as the old Silurian is concerned, and this is the tale.

" When poor De la Beche was gone, very old and valued geological friends (not of the Survey) urged me to look to the place as the man who, by his labours in British fields, and his application of his knowledge to maps, sections, and

books, was most entitled to the post, and who, from successful management of Societies, could best succeed in it. But
I peremptorily declined not only this appeal, but also a
gentle allusion of the President of the Board of Trade when
he sounded me *in limine*. At that time my great fear was,
that geology would be submerged in other affairs if a good
hammer-man was not at the head of the whole. , , .

"Who, then, was to be (Geology and Palæontology apart)
the *régisseur ?* A split was to be deprecated—two kings
would never answer, and confusion would arise. Yesterday
sundry Professors, the four or five who are the oldest and most
influential, met together in Jermyn Street and unanimously
opined that I was the only man who could keep the whole
thing together and make it work well. This opinion they
conveyed to the authorities through Playfair, and the appeal
being made to me in so very flattering a way, I could not
resist, particularly as I saw that I should, by accepting,
prevent discord. Whether the Government will offer it to
me is another question,

"Notwithstanding your *mot* on the triple directorate, I
view it simply as the School of British Geology and Mines,
The affiliated sciences are all subordinate to that fundamental point.

"If they name me, and should my health continue as
good as now, you know me well enough to be certain that I
will do some good at all events, though perhaps I have
offered to undertake what I had better have left alone, as far
as my happiness is concerned.—Yours most truly,

"ROD. I. MURCHISON."

"30*th April*.—MY DEAR SEDGWICK,—I was just going to
write to you to thank you with all heartiness for your kind-

ness in adding your name to the list of good men who recommended me to the Government to succeed poor De la Beche. I would rather have had your name than that of any living man, and it quite rejoiced me to see it.

" The list when sent in to the Government was really most gratifying to me, for it included the names of almost every man of real mark in various sciences, from the President of the Royal Society downwards, and so your visit to the Geologicals and Royals, if it *pro tanto* injured your health, was of real service to your old friend.

" But though this potent list, backed by the unanimous wish of the Professors in Jermyn Street (who did not sign, they being employés of the Government), was sent in a week ago, it was only yesterday that Lord Palmerston sent for me ; and, after receiving a good cram ' *de rebus geologicis,*' said he would beg the Queen to appoint me, and would accompany his letter to Her Majesty with the recommendations of my friends. So old Fitton's scrawl and your name are now before Her Majesty, and the thing may be considered done, though of course I can do nothing in Jermyn Street, until I receive due notification thereof."[1]

In a letter to Phillips of the same date he writes :—

" I am in hopes (God willing) of being of some use, and more particularly in diffusing geological knowledge more effectively through our upper classes, who are of all the community the worst informed in our lore. I will back many a mechanic attending the evening lectures in Jermyn Street

[1] With reference to this recommendatory memorial Professor Ramsay remarks :—" I proposed to the Professors that Murchison should be Director-General. They agreed. Mr. Cardwell was written to on the subject, and as he approved, Sir Roderick was then spoken to. The appointment seems to have been practically settled before "Fitton's scrawl" went before Her Majesty.

against most of our senators.   Many many thanks for your
true friendliness.   Jukes writes to me from Ireland in delight at the prospect of what was announced to him as the
wish of the good men in Jermyn Street.   I never would
have thought of it, unless unanimity existed.   The moment
my friends prevailed over my 'nolo episcopari,' there was
but one voice, and I am most grateful for it; indeed, I value
that expression of good-will more than the place or any gift
of the Crown."

It was a somewhat hazardous experiment for a man in
his sixty-fourth year, who for nearly forty years had been
free to go where he pleased, and who had rejoiced to take
the full use of his freedom, to subject himself now to the
trammels of an office requiring for its effectual performance
constant attendance and watchfulness.   He begins as might
have been expected of him.   The very first day he discovers
that there is no proper general catalogue of the contents of
the museum, and he forthwith gives directions for the call-
ing of a council of Professors to prepare a report on the subject of a catalogue.   When he entered on his new office, he
daily recorded, in a sort of official journal, the items of business he had transacted,—a practice, however, which soon
fell into disuse.   From one of the early pages of his diary
an extract of more general interest may be taken :—

" *May* 11*th*.[1]—Interview with Prince Albert at Buckingham Palace.   Was complimented by H.R.H. on my appointment.   Explained to him some of my views, some of
our *desiderata,* and some of our doings ; particularly called
his attention to poor De la Beche's last *Catalogue of British
Pottery and Porcelain,* with which he seemed well pleased.

[1] He had entered on his duties on 5th May.

" The Prince then explained all his views as to the hope of realizing, at a future day, a concentration of all the chief scientific, artistic, and manufactured produce in one great building, and how the possibility of such an advance was stopped by the want of money and the unwillingness of Government to spend in these warlike times : suggested a modified scheme for the erection of a temporary building of corrugated iron with glass, slightly raised above the ground, and warmed by hot water, whereby specimens now spoiling in damp places might be kept :[1] hoped that £2000 or £3000 would be granted for such a purpose : regretted that the frequent changes of Government brought new men continually on the *tapis*, and nothing permanent was carried on.

" He gave a pretty little lecture on the desirableness of having at hand and united in the same suite of buildings professors who could illustrate every part of a porcelain specimen, *e.g.* :—

1. As to the nature of the materials.
2. As to the chemical changes produced by fire, and mixture of ingredients.
3. As to the methods of painting, enamelling, and embossing.
4. As to archæology, date, etc.

H.R.H. then alluded to an influence now at work to have the Royal Academy transferred to the new building at Burlington House, and regretted that the men of science

---

[1] This project points to the future buildings erected on the plan here stated, and known popularly, from their site and shape, as " the Brompton Boilers." In due course, as the Prince hoped, they have given place to the handsome buildings in which the treasures of the South Kensington Museum are now housed.

should not be there. I replied that we men of science should put our shoulders to the wheel, and endeavour to keep the Government to the proposed occupation by ourselves." [1]

A natural consequence of the accession of official dignity was an augmented display of hospitality at No. 16 Belgrave Square. Writing, for instance to Sedgwick, the new Director-General expresses himself thus :—" Besides the enclosed ' bow, tea, negus, ice, and turnouts,' I ask my special friends who testified for me as fit to direct the Geological Survey to eat whitebait with me at Greenwich on Wednesday the 20th. I hope you will be able to come."

In spite, however, of business and festivity, his first trial of official work seems to have pleased him, for after a few weeks of experience he entered in his now irregularly kept office-diary the following rather formidable syllabus of work already accomplished :—

" *July* 16*th.*—Nearly three months have elapsed since I began to methodize and record my daily work. I soon found this to be impracticable. I have worked here daily from 11 to 4 and half-past 4, sometimes from half-past 10 to 5, and even near to 6 o'clock. I do not find the work to disagree with me. On the contrary, I am in better health than when I began, notwithstanding the excitement of lecturing to crowds on the top of Malvern Hills,[2] my great Greenwich fête to forty savans, and all the dinners and parties, public and private.

" 1. I have now pretty nearly arranged a new and uniform

[1] By the recent remodelling and enlargement of Burlington House, space has been given there for the Royal Academy and for the chartered learned Societies.

[2] There had been a meeting of naturalists, at which Murchison was present, and discoursed on his favourite science from the top of the Worcestershire Beacon.

system of colouring the maps by which each natural group will be represented by one base colour, and the subdivisions of the same by varied tints thereof.

" 2. I have ordered the lettering of all the colours on the maps.

" 3. I have employed Morris, and commenced a rigorous purification of the contents of the fossil department.

" 4. I have drawn out and designed the duties of the Curator of the Museum and his assistant, and those of a Palæontologist, or, as I hope, Palæontologists.

" 5. House secured for the use of Dr. Percy.

" 6. Strong appeals have been made to those on the Board of Works to enable us to get our new rooms adjacent to this building.

" 7. I organized for Lord Clarendon the expedition to the Gulf of Nicomedia in search of coal, and obtained the appointment of Mr. H. Poole as surveyor.

" 8. I obtained from Lord Clarendon, to send to the American Geographical Society of New York, the volumes of the British Association and Statistical Society, to which were added our own volumes.

" 9. I have rendered the titles of all our future volumes uniform,—Records of Mines, Decades, Memoirs are all to appear under the general title of *Memoirs of the Geological Survey.*

" 10. I have applied to have the officers and assistants of the Museum placed on the same footing as those of Marlborough House,[1] in obtaining a regular holiday of one month, viz., from 10th August to 10th September.

[1] The offices of the Department of Science and Art under which the Geological Survey was placed.

" 11. On entering office I made a vigorous stand against
a Parliamentary document, drawn up by Playfair as Secretary of the Department at the Board of Trade, whereby
Mr. Cole was named ' Inspector-General ' of all schools and
Museums, whether in the metropolis or country.   I insisted
on a special exemption of this establishment from such
a rule, and a paragraph to that effect was accordingly inserted.

" 12. Seeing the accumulation of volumes of the Survey
publications, I have already begun to distribute the extra
copies (preserving always a good reserve) to foreign bodies
which will really appreciate them.   When the list is properly prepared, such a distribution will be highly useful in
making our services widely known, and their value recognised throughout the world.   Hitherto the results of the
very valuable labours of my predecessor and his associates
have not been by any means sufficiently diffused.

" These affairs, as above cursorily noted, to say nothing
of a thousand details, correspondences, boards, council-meet-
ings, and so forth, have completely occupied my first ten
weeks of office.—R. I. M."

Among the correspondences here referred to, he still kept
up the old friendly one with Sedgwick, to whom, on the
30th of May, he writes as follows :—" I am sorry to hear of
your ailments.   I trust that you are much too foreboding
respecting your duration of life.   You are no older than our
Prime Minister, who has to face angry Houses of Parliament
nightly, and is never in bed till one or two in the morning.

" Your *P.S.* announcing your third *fasciculus* or introduction, shows that you are as active in mind as ever.
Quite agreeing with you that half-measures in arguments in

science are no measures, I would very much regret if you
fire any such Minié rifle shots at your old friend as require
to be answered except in a perfectly friendly manner.

" By the bye, you have no doubt heard of (if I did not
already tell you of) the discoveries of fossils in our Durness
limestone of Sutherland, by Peach. He has corresponded
with me on the point, and has sent me some of the fossils.
I have had them polished. The forms (rude and ill preserved as they are) look more like *Clymeniœ* and *Goniatites*
than anything else (with corals) ; and if so, the calcareous
masses which we saw from Assynt to Durness, interstratified
in the quartz rock, are high in the Devonian ! I would
like to hear what you say to this *éclaircissement.* I see
great difficulty in understanding it.

" If the conglomerates of the Ord of Caithness and Ben
Bhragie, close to Dunrobin, are the equivalents of the West
Sutherland quartz rocks, they must also be so of the Scarabin
hills, which are in contact with the true Old Red of the east
coast. If, on the contrary, these crystalline rocks should
prove altered equivalents of Silurian strata, I see nothing but
what is rational.

" It is twenty-eight years since we tramped across
Sutherland, and the going over of my well-kept journal (in
which I have some of your writing and much of your mind)
has been a source of great pleasure to me, ruffled only by
your announcement of the forthcoming continuation of our
disputation on things of which neither of us had an inkling
in 1827."

But besides reviving such pleasant memories, the discovery of fossils in these rocks of the far north awakened in
the veteran geologist a strong desire to revisit the ground,

and the desire soon passed into a settled determination. Evidently a most important problem in British geology was to be worked out, and one, moreover, to which, by previous examination of the ground, he perhaps possessed a clue. Accordingly he arranged to start once more for a survey of the rocks of the Highlands. The British Association was to hold its meeting this year in Glasgow, and he could combine attendance there with his northern tour. This excursion to the north-western headlands of Scotland proved to be the beginning of a series of Highland tours, extending over a period of five years, during which Murchison, in concert with different fellow-labourers, accomplished the last great scientific achievement of his life.

On the 7th August, the day before setting out for the north, the following entry occurs in the office-diary :—

" Last day in the office. Before I leave, I am glad to have made two good moves among, I hope, many others : the one applying to my friend Lord Canning, as Governor-General of India, and begging him to look to the geological structure of India, and have surveyors in all the Presidencies ; the other to Sir W. Molesworth, the new Secretary for the Colonies, urging him to do the same in many of our neglected Colonies. The answers from both are favourable, and I have hopes of something better.

" On starting for the Highlands, I may fairly say that the occupation I have had here has benefited me, and realized the opinion of Cicero, to which old Dr. Fowler, of Salisbury, adverted,—' Manent ingenia senibus, modo permaneant studia et industria.' " [1]

Thus busily and pleasantly passed away the first few

[1] See *Brit. Assoc. Reports*, 1854, Part ii. p. 114.

months of official duty. From this time onward, though it
may need only occasionally special notice in our narrative,
the work of the Geological Survey continued to be Murchison's chief employment. It did not by any means engross
all his time, for the vigour with which he had entered into
the duties of his office soon toned down. He had leisure
for a summer ramble, and we now enter on the details of
the first of a series of such rambles in his native Highlands. He could likewise still devote himself to the interests of the Geographical Society, and take part, as of old,
in the business of the Geological. But beneath all these
more prominent avocations there lay a constant under-
current of official duty which, even when it might be merely
of a routine kind, necessarily demanded for its due performance a good deal of time.

CHAPTER XXIII.

IN the early days of geological speculation, it was a favourite belief among the disciples of Werner that the primeval shoreless ocean which tumbled round the globe held in its hot and fecund waters the substance of all rocks, and that the earliest of the deposits which settled down on its floor were such as have received the names of gneiss and schist. The followers of Hutton, on the other hand, maintained that these rocks do not exist now in their original condition, that at first they were mere ordinary sediment, like the mud, silt, and sand of the present sea-bed, and that they have subsequently been squeezed, hardened, and rendered crystalline by the action of the earth's internal heat. With the advance of years, the essential truth of the Huttonian creed became clear. All over the world proofs were obtained that gneiss, schist, and similar rocks, instead of being necessarily the oldest formations, belonged to many different geological ages. Sometimes the sedimentary deposits of one period, sometimes those of another, had come within the influence of underground movement and heat, with the result of assuming a more or less perfectly crystalline character. Hence, without a careful scrutiny of the surrounding region, it was scarcely possible to decide as to the geological date of any mass of such crystalline rocks. Nevertheless, the influence of the old creed continued to show itself. Gneiss and schist, about whose relative date nothing definite was known, were still put down below the fossiliferous formations as " primary " rocks, lying almost beyond the furthest limits of geological history.

No better illustration of this condition of things could be furnished than that presented by the crystalline masses which form the Highlands of Scotland. Looking back from our present knowledge to the state in which these rocks were allowed to lie for more than half a century, it seems as if, with a perverse ingenuity, geologists had refused to examine them, although all the while in possession of the clew to their history. The rocks still appeared in geological maps and text-books as ancient formations lying beneath all the fossiliferous systems, and even as " azoic," or earlier than the introduction of life into the earth.

In the earlier years of this century, Macculloch had made some important observations on the geology of that wild mountainous region which stretches from the Kyles of Skye northwards to Cape Wrath. He was sorely puzzled by the red sandstones which rise into the huge terraced hills of Applecross and Assynt. By any one coming to them fresh from an Old Red Sandstone region, they would be set down unhesitatingly as belonging to that geological system. But Macculloch, whose first belief was that such should be the position assigned to them, found that they passed beneath quartz-rock and other members of the so-called " primary " series. Hence he classed them as primary red sandstones,

though how they came to be there, and what were their
equivalents elsewhere, he felt himself wholly at a loss to
decide.   In the course of his rambles among the quartz-
rocks he had observed many curious tubular objects in those
rocks, and surmised that they had an organic origin.   Fossils
in the " primary rocks " !   It could not be.   By common consent the " primary rocks " were regarded as dating from a time
earlier than the appearance of life upon the globe—a time of
heat and turmoil, and chemical precipitation from a thermal
ocean, when no life was possible.   Such being the prevalent
notions, Macculloch's remarkable observations excited little
interest.   Probably he did not even himself perceive their
value and bearing.   And yet, regarding them in the light

Worm-burrows in the Quartz-rock of the North-west Highlands.

of what is now known of the geology of the Highlands, we
cannot but perceive that could any active geologist, not too
enthralled by theory, have gone into that northern region
determined to work out the whole problem, he might have
rendered a very great service to geology by introducing true
notions regarding the origin and position of the so-called
metamorphic rocks many years before they were otherwise
slowly and painfully worked out.

No such fortunate explorer, however, had sought these
wilds.   In the year 1827, Sedgwick and Murchison paid the
visit to them narrated in the eighth Chapter of this biography.
But their minds were so full of Old Red Sandstone at the
time that they failed to realize the nature of the difficulties

which Macculloch had encountered. They quietly set aside his observations with the remark, that they thought it absolutely impossible to separate the red sandstone of Sutherland from the admitted Old Red Sandstone of Caithness, and that they must all be " classed with the older secondary deposits of England." With equal curtness they dismiss his supposed fossil shells—" we cannot regard them as organic."

In the year 1840, R. J. H. Cunningham—a man too soon lost to science—ventured once more into the fastnesses of Sutherland, with the view of ascertaining something about their geology. Getting away westward into the rugged ground of Assynt, he confirmed and extended the observations of Macculloch. In a singularly able Memoir, which unfortunately, from the mode of its publication, has been comparatively little known to geologists,[1] he showed that the red sandstones lie upon an old gneiss, and pass under quartz-rocks and limestones, over which comes an upper gneiss. He completely corroborated the remark of the earlier geologist as to the organic nature of the bodies in the quartz-rock, and found them in abundance *in situ*, though Sedgwick and Murchison said they could pick them up only in loose blocks of stone. And we find him shrewdly remarking on the fact, then new and unexpected, that there are gneisses and mica-schists not of the oldest antiquity, but actually later in origin than the creation of organized beings. Nobody seems to have thought it a desirable thing to find out more about these organisms, now proved to exist beneath a great mass of the ancient crystalline rocks of the Highlands. Nay, the very existence of such fossils seems to have passed

---

[1] It was published as a Prize Essay in the 13th volume of the *Transactions of the Highland Society.*

out of sight, and for fifteen years later the gneisses and
schists of Scotland continued to be classed with the primeval
masses which were supposed to be earlier in date than the
first beginnings of life.

How much longer the ignorance and indifference of
Scottish geologists might have lasted it may be impossible
to say, had not a happy accident once more called attention
to the forgotten fossils of Macculloch and Cunningham.   In
the year 1854, Mr. C. W. Peach, who had long been known
as one of the most keen-eyed collectors in Britain, and who,
moreover, added to his other powers an excellent knowledge
of natural history, so that he knew the value and meaning
of what he found, had occasion to leave Wick, where he
was stationed as one of the officers of Customs, and visit a
wrecked ship on the coast of Sutherland.   Looking over
some of the weathered blocks of the limestones of Durness,
he found certain bodies which, though imperfect, were unquestionably shells.   These were sent by him to Murchison,
who, as we have seen, at once perceived the important
bearing which they might have in settling the age of the
rocks of the Highlands.   At first they were compared to
some Devonian forms, but, as the latter geologist remarked
at the time, this could hardly be their true position, unless
Sedgwick and he had made some egregious blunder in their
sections of that north-west region of Scotland.   These two
explorers had indeed been mistaken in their sections, but not
quite so much as this identification of the fossils with
Devonian species would have shown.   They had correctly
placed all the Highland crystalline rocks below the Old Red
Sandstone.

It was to clear up the question of the true horizon of the

fossils that Murchison closed his diary at Jermyn Street and set his face once more towards the north. He had not left himself much time, for it was now the end of the first week in August (1855), and the British Association was to meet in Glasgow on the 12th of September, and he had promised to be present. But the point to be settled lay within such a narrow compass, that, with any tolerable weather, he might reasonably hope to succeed, and to bring back the announcement of an important new step in British geology.

With Professor Nicol as his companion, Murchison sped north by the Caledonian Canal to Inverness—not the shortest route for the headlands of the north-west of Sutherland, but one which gave him another peep at the scenes of his infancy. It was a bright autumnal morning as they skirted the green shores of the Beauly Firth. The old Tower of Fairburn caught the sunlight as they passed, and Tarradale looked so quiet and peaceful, nestling down among its old trees, that the former feeling of regret came back again that it had been sold. Time had been busy even in that retired spot. Among other changes, it was found that the "mailers" or cottagers, who had formed so conspicuous a feature of the district in early days, were all gone. The little huts and rude stone walls had been cleared off to make way for the large well-tilled fields, which now recalled the advanced agriculture of the Lothians.

From the thoughts which these changes suggested, the transition was easy to those called up by the old burying-ground of the Fairburn family at Dingwall, about which Murchison at once proceeded to inquire. His uncle the General had with pious care repaired the family tomb, where, among other maternal ancestors, lay the Rory More or

Big Rory Mackenzie, before mentioned.   The effigy of that
stalwart Highlander, carved in stone, his prosaic descendant
now proceeded deliberately to measure, entering in his
note-book, " 6 feet 4 inches."

Amid storms of wind and rain congenial to that bois-
terous climate, the travellers drove through Sutherland, and
at last broke ground in Assynt   Eight-and-twenty years
had passed since that stormy season when the future
knights of Cambria and Siluria, then in the full vigour and
merriment of their early prime, boated and climbed through
these northern wilds.   Murchison had never been back
during the long interval.   But now, partly from the dreadful
weather, and partly from finding so many of his old acquaintances gone to their rest, Assynt wore to him a gloomy
and depressing aspect.   He writes, " I found the old house
at Balnakill new roofing; old Dunlop dead ; my kind old
friend, Anderson of Rispond, who sheltered Walter Scott as
well as Sedgwick and self, dead.   All around gave note that
my day was fast coming, and that I had taken my farewell
look at the Whiten and Far-out Heads."   Brighter skies
however, eventually dispelled the gloom, and he lived to
come back again, and yet again, to those headlands.

The excursion was too brief to permit of much good
field-work being done in so rough a country.   In spite of
the clear statements of Cunningham, the travellers could
not shake off the old prejudice in favour of the Old Red
Sandstone age of the vast red conglomerates and sandstones
of Assynt and Applecross.   They saw indeed that a red conglomerate and grit did lie between the quartz-rocks and the
older gneiss, and could not but admit that some at least of the
red rocks must be older than the quartz-rocks and lime-

stones. But the influence of an old and erroneous observation made in the tour in 1827 kept Murchison from understanding the full meaning of what he saw, and from drawing
the natural inferences from sections which he sketched correctly in his note-book.[1] He still regarded it as clear that
the stupendous masses of sandstone in Gareloch and Applecross must be classed with the Old Red Sandstone, so
that as far as respected any new light on the geology of the
north-west of Scotland his excursion to Assynt left matters
very much where they were. He reiterated his belief that
the quartz-rocks, limestones, and over-lying gneiss and schist
could not possibly belong to the Old Red Sandstone, but
must be of much older, perhaps of Silurian, date.

That the results of this tour were not very convincing
is further shown by the fact that Professor Nicol felt so little
impressed by Murchison's reasoning as to the high antiquity
of these crystalline rocks, that he seriously suggested that
they might even prove to be of Carboniferous date. Evidently field-geology was at fault in the meantime, and the
first strong ray of light would be thrown on the obscurity
of the whole question by the discovery of fossils sufficiently
well marked to place their geological horizon beyond all
doubt. This discovery was made before long by Mr.
Peach.

The return journey included a visit to Wick and the
wonderful coast cliffs of the east of Caithness; the granite

---

[1] It is remarkable that in faithfully drawing in his note-book what
he saw in nature, Murchison correctly represented in several sections the
unconformability of the quartz-rocks on the red sandstones. Yet he
missed the meaning of this fundamental feature of the geology, and took
no notice of it at the time. It was re-observed, and its value was fully
realized, next year by Sir Henry James and Professor Nicol, as will be
told in later pages.

and conglomerate of the Ord; Brora, with its moraines, which Murchison refused to believe to be the work of terrestrial glaciers; Dunrobin, with its hospitable and public-spirited Duke and Duchess, and its reefs of Jurassic strata; Ardross, and the Alness sections of the Old Red Sandstone and other attractions, which could receive but a rapid glance, for the Association week was now close at hand.

At the Glasgow meeting of the British Association in 1855 the geologists mustered strongly. Murchison was again chosen as President of Section C. Sedgwick, shaking off the thousand and one cares or ailments which usually impeded his action, came up once more among his northern brethren of the hammer. He had just issued the Introduction to his *Synopsis of British Palæozoic Fossils,* in which he had spoken of his former associate Murchison with a vehemence of language very characteristic, but regretted by all who admired and respected both disputants. To any reader who had no personal knowledge of them, Sedgwick's tirade must have seemed expressive of bitter animosity and estrangement. Probably he was not unaware of this himself. For when, after Murchison's communication on the Sutherland-shire story, he rose to speak, and began by deliberately taking off a thick heavy great-coat in which he had been sitting, he noticed that he had raised a smile in the audience, where-upon he instantly and in his happiest way exclaimed, " Oh, I 'm not going to fight him ! " The smile passed at once into a good laugh and general applause through the room. There had been some discussion on the true place of the crystal-line rocks of the Highlands overlying those quartz-rocks and limestones of Sutherlandshire which had yielded fossils. Hugh Miller, with his rough shepherd's plaid and his shaggy sandy hair, talked as he wrote in vigorous English, and with his strong north-country accent maintained that the rocks in question were metamorphosed portions of his own Old Red Sandstone. Against this contention Murchison strongly pro- tested, appealing to his companion Nicol, as well as his old associate of 1827, in support of the fundamental fact that whatever might be the age of the rocks, they were at least vastly older than the Old Red Sandstone. Sedgwick, divested of his coat, plunged at once into the debate, and soon branched off into endless humorous episodes and digressions. He had occasion, for instance, to mention some natural section of rocks much overgrown with wood, on which, abruptly stop- ping his geological disquisition, and as it were taking the audience into his confidence, he quietly remarked, " By the way, ladies and gentlemen, trees are a confounded impedi- ment to the progress of geological research." He ended by taking the same view as Murchison with regard to the relative antiquity of the limestones and the red conglo- merates. But it was clear that the rocks of the north-west of Scotland still presented a very curious and interesting problem, which could not be solved without more and better fossils, and further extended examination of the ground. To this renewed research Murchison resolved to devote himself.

At this gathering of the British Association Professor Ramsay was able to announce the beginning and first year's progress of the Geological Survey in Scotland. He himself had started the work on the coast of East Lothian, and some advance had been made in tracing the areas of Old Red Sandstone, Carboniferous and Silurian rocks, on the large six-inch Ordnance maps, which were exhibited in the Geolo- gical Section.

Another question of considerable interest in Silurian geology arose at the meeting. From the moory uplands of Lanarkshire—the old haunts and hiding-places of the Covenanters—there had come, for the inspection of the learned, a series of remarkable fossils, collected in the course of his

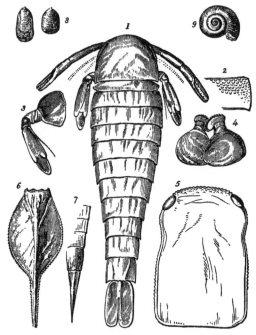

CRUSTACEANS AND SHELLS OF THE UPPER SILURIAN ROCKS OF LANARKSHIRE.

1. Pterygotus bilobus, half natural size [in this restoration the body is rather too narrow, and has one segment too few]. 2. Portion of a body-joint, to show the peculiar sculpture. 3. Swimming-foot, nat. size. 4. Pair of foliaceous joints. 5. Slimonia acuminata, head, half nat. size. 6. Tail of same, one-fourth nat. size. 7. Tail and some body-joints of Eurypterus lanceolatus. 8. Lingula cornea. 9. Platychisma helicites.

long professional walks and few spare leisure hours by Mr. R. Slimon, a country doctor at Lesmahagow. A cursory examination of these specimens showed that they were a fine series of well-preserved lobster-like crustaceans, apparently identical with forms found in the Upper Silurian and

Lower Old Red Sandstone. If it could be shown that the strata yielding them really belonged to the Upper Silurian, and if their position and their relations to other formations could be fixed, then a new and important step would be made in establishing the Silurian system on the north of the Tweed. As soon as the meeting was over Murchison started for the Lanarkshire ground, accompanied by his friend and colleague Professor Ramsay. The general results of this rapid raid were thus told to Professor Harkness :—

" I came last night direct from the Lesmahagow country, where I passed two entire days with the good Slimon, and took Ramsay with me. We had glorious weather. I am more satisfied with my general results than anything I have seen for many a day. . . .

"The merit of this discovery is Slimon's, and I shall raise a statue to him for it. As I had a very fast horse to take me from point to point, and used my legs up every burn (including all the Carboniferous series to the east), and to the top of Nutberry, and considerably to the west of it, I have no doubt as to the completeness of the evidence. . . .

" The Old Red is a poor affair there,[1] one member of it, *i.e.* the Lower, is undoubted, and great dislocations separate it from the Lower Carboniferous. But the gradation into Upper Silurian on the Logan Water, and the alternation of red and grey until all becomes grey, is the best exemplification of transition that can be. In short, it is equal to any passages of the same rocks in Hereford, Salop, and Cumberland.

[1] This was an error. The Lower Old Red Sandstone alone measures in that district at least from 10,000 to 15,000 feet.

" When Salter has examined the fossils I will tell you more. In the meantime I have left Slimon the happiest man possible, and I intend, *D. V.*, to give a little introduction to the description of the wonderful geological parish of Lesmahagow, and the merits of the poor but meritorious Dr. Slimon, who, if he had been rich enough to visit his patients on a horse, and had not travelled up the braes on foot, would never have made this excellent hit. So, what with my crystallized Lower Silurian[1] fossils in quartz-rock and marble of the Highlands, and the true uppermost Silurians of the south of Scotland, I consider this a capital summer."

And a very capital summer it proved to be. That brilliant dash at the Lanarkshire ground brought to light for the first time in Scotland the true base of the Old Red Sandstone, and the true top of the Silurian formations. It thus gave a new starting-point for the further investigation of the geology of the country, and for the comparison of the rocks of Scotland with those of England and Wales. Still more important, however, was the step taken with regard to the structure of the north-west Highlands. It was the first of a series of excursions which revealed the actual foundation stones on which all the geological formations of Britain have been built, and brought the so-called azoic schists and gneisses of Scotland into relation with the fossiliferous Silurian deposits of other regions. The excursions, however, were not continuously prosecuted in successive years. Murchison had at once recognised the importance of the subject.

---

[1] It had not yet been decisively proved that the fossils were Lower Silurian, though the chances of their turning out to belong to that horizon were very great, so great that Murchison regarded the point as practically settled.

But the materials, as we have seen, did not at first exist for the complete solution of the problem. Before they were obtained other questions arose, and the strong tide of life and work in London swept him elsewhere than to his beloved Highlands.

CHAPTER XXIV.

For the sake of continuous narration, we must now depart from chronological order, and pass over an interval of more than two years, so as to follow without interruption the steps whereby the Highland problem was gradually worked out.

As the summer of 1858 wore round, and London began to empty, it was time to prepare for the usual autumnal holiday. During the years which had intervened since the ramble into Sutherlandshire to see the whereabouts of the fossil-bearing rocks of Durness, while Murchison was spending his vacation on the Continent, Mr. Peach had returned to the Highland ground, and had found other and better fossils, which placed beyond dispute the Lower Silurian age of the quartz-rocks and limestones of the north-west Highlands. Other observers had likewise been at work. Professor Nicol and Colonel Sir Henry James (now the Director of the Ordnance Survey), having independently traced the boundaries of the formations for long distances, had shown by fresh proofs the infraposition of the great red sandstones and conglomerates to the quartz-rocks; discovering, at the same time, not only that the quartz-rocks lay above the red sandstones, but that they did so with a strong unconformability, there being thus a complete break and discordance between the two formations.[1] Mr. Nicol had likewise traced the regular succession of quartz-rocks and limestones passing under the vast overlying mass of flaggy gneiss, which is the uppermost rock of the district. Murchison determined to spend his autumn this year in trying to make out more precisely the relations of these Highland rocks to each other and to the rest of the geology of the country. The Old Red Sandstone of those northern tracts had never been very satisfactorily put into order by any geologist, and as he had an offer of a place in the steamer of the Commissioners of Northern Lights on its annual voyage among the Orkney and Shetland Islands, where the Old Red Sandstone is largely developed, he arranged to include these remote regions in his programme. He now never undertook any geological journey without a companion. This year he chose Mr. Peach, whose keen eye and personal acquaintance with a good deal of the ground to be visited would doubtless be of much service.

In the early part of the journey we find the two geologists, hammer in hand, among the flagstone quarries of Thurso, gathering some of the fossil fishes so characteristic of the rocks there, or in the shop of the baker, botanist, geologist, and poet, Robert Dick, getting from him a sketch of the distribution of the rocks, which he graphically depicted with flour on one of his own baking-boards. Taking the packet from Thurso for Stromness, they could see—what must strike the eye of every traveller to that hyperborean tract—the contrast in

---

[1] Colonel James communicated his observations in a letter to Murchison, dated 26th July 1856. Mr. Nicol's were made independently a little later in the same summer.

form and colour between the two great horns or headlands of
the noble bay. On one side the dark solid flagstone cliffs
of Holburn Head rise boldly out of the sea, cut into square
masses by deep narrow clefts, into which the surge is ever
rolling, pierced too with sea-caves and fronted by square
gaunt buttress-like columns of the same dark stone, edgéd
here and there with bright lichen, and sharply trenched
with lines of inky shadow. On the opposite side the rocks
assume a warm rich tone of colour, which seems at first
more suggestive of a sunny Italian landscape than of the
bleak sombre north. Barred with lines of red and yellow,
and varied blending tints of green and brown, the rocks rise
in shattered ruinous cliffs high above the breakers, which in
calm weather play about their base, and in storms cover
them with sheets of foam. The geologist who fixes well in
his mind these two types of cliffs, and by visiting them and
realizing on the spot the character of the rocks on which the
contrast depends, carries with him the key to some of the
most interesting geology farther north.

It was after having seen the rocks of each type that
Murchison and his companion turned towards Orkney.
Passing under the grand precipices of Hoy, they recognised
the reappearance of the rocks of the east side of Thurso bay,
and turning round the end of that island they found themselves, in approaching Stromness, among the same flagstones
as at Holburn Head. It was from these flagstones that Hugh
Miller had obtained the *Asterolepis* he so graphically described,
and on whose anatomy he hung his disquisition against the
Vestiges of Creation. From quaint Norse-like Stromness the
road leads by the Lake of Stennis, which had likewise furnished Miller with some apt analogies, past the ghost-like

circle of Standing Stones to Kirkwall. It was at this Orcadian capital that our travellers were to await the 'Pharos'
steamer. They meanwhile employed the time in geological
exploration, noting among other things the traces of land
plants and great abundance of fossil fishes in the rocks, and
suggesting that the island of Pomona might consequently
be aptly termed *Piscina*. The 'Pharos' duly appeared, took
her two guests on board, and, turning her head northwards,

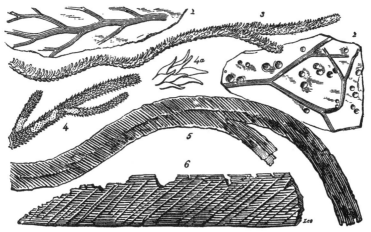

LAND PLANTS OF THE OLD RED SANDSTONE OF CAITHNESS.

1. Branched rootlets of some (Lycopodiaceous?) plant. 2. roots of Lepidodendron (?)
with double annelide-burrows. 3. Lycopodites Milleri, one-third natural size. 4. Lepidodendron nothum. 5. Flattened root, and 6. fluted stem, of coniferous tree.

was soon ploughing the rough seas which wash the northern-
most of the Orkney Islands.

On the morning of the 4th of August the noble cliffs
of Sumburgh Head, the most southern point of Shetland,
stood up in front, their ledges white with sea-fowl, their base
fretted with surge, and enlivened by the boats of the bold
fisher-folk. A geological eye could recognise in these cliffs
the same features which mark the headlands to the west of

Thurso, and which stretch through the islands of Orkney. But away a few miles to the north-west, old Norna's precipices of the Fitful Head looked out grandly upon the wide Atlantic, and showed in their form and colour a change from the type of rock so characteristic of the sea-coast of Caithness and Orkney. The voyage of the 'Pharos,' however, being for the definite purpose of visiting the lighthouses, comparatively little scope was afforded to the hammermen. They had a chance now and then of landing at the different stations touched at on the way, and for the rest had to content themselves with straining their eyes to make the most of such geological indications as could be deciphered, by means of field-glasses, in the cliffs and shore-ledges as they passed along. They saw and found enough, however, to convince them that the Old Red Sandstone is prolonged into the southern limb of the Shetland Islands.

But in this pleasant cruise, where, in spite of their efforts, the geologists found that science could play but a subordinate part, by far the most striking lesson learnt was that taught by some of the little islets or skerries on the east side of the group not far from Lerwick. Even in such calm weather as that which the 'Pharos' now enjoyed, the ceaseless curling and cresting of the surf beneath the cliffs was enough to remind one that the scene must be very different during a storm. Placed among some of the most rapid and conflicting tideways, and exposed to the unchecked fury of the winds and waves, these islands furnish probably the most appalling proofs anywhere to be met with in the British seas of the destructive power of the ocean. When the breakers are at their worst, vast sheets of water, rising to 100 feet or more above the ordinary level of the sea, are driven over the rocks and skerries. Blocks of rock, many tons in weight, are thus actually quarried out of their solid beds, at heights of sixty and seventy feet above high-water mark, and driven along the rugged ground, grinding and scoring the rocks over which they are borne. Piles of huge angular blocks are thus accumulated so high above the common limit of the waves, that a believer in ancient convulsions, like Murchison, might well have been pardoned for citing them as evidence of the far higher intensity of former geological operations. Yet that geologist, under the able guidance of Mr. Thomas Stevenson, the engineer, who had given special attention to the subject, came away thoroughly convinced that the apparent traces of primeval havoc were due to the present action of the sea, and that some portions of the ruin had been piled up only a winter or two before.[1]

The limit of the cruise was fixed by the last lighthouse to be visited—that on one of the little islets in front of the farthest headland of Shetland. This forms the most northerly part of the British Isles. Murchison had never been so near the North Pole before, so he duly inscribed his name in the lighthouse books. In the adjacent island of Unst he met with Mr. T. Edmonstone, and from him got the story of his interview with the Duke of Wellington, who on the occasion of the Queen's coronation received the Shetlander kindly, took him to the coronation-ball, and then to spend a day or two at Strathfieldsaye. "The Duke asked him how he had been amused at the fête, and what struck him most, whereupon old Ed-

---

[1] Mr. Stevenson has given a careful description of the phenomena as exhibited in the Bound Skerry of Whalsey.—See his work on *Harboars*, p. 22.

monstone replied, 'To watch the meeting between your
Grace and Marshal Soult, whom I was told you had never
met except in the field.'   'That is not quite true,' said
the Duke, 'for after the capture of Paris, and when we
were all much tired, I, who had not had my clothes off for
two or three nights, was deputed to go out to meet Louis
the Eighteenth. In the carriage I fell fast asleep, and at
the first relay Soult came up, and, on inquiry, looked at me
sleeping and passed on. It was the first time he ever
caught me napping!'"

From the Ultima Thule of these northern isles the
'Pharos' steamed southwards again, and finally dropped the
geologists on the bleak headland of Cape Wrath. From that
point, left once more to their own devices on the solid earth,
they came eastward and southward, revisiting the old sections of 1827, as well as those of 1855, and making new
traverses, so as to get a clear notion of the arrangement of
the rocks. They found that order to be as recently stated
by Professor Nicol and Colonel James. It was now clear
that the thick masses of red sandstone, in spite of their
usually gentle inclination, were older than the quartz-rocks
and limestones, and that these in turn were older than the
schist and gneiss which, overlying them and undulating
away towards the east, formed the greater part of the mountains of Sutherland. The fossils found by Mr. Peach had
not only shown the rocks to be Lower Silurian, but to present a curious analogy to some portions of the corresponding series in North America. In all, nineteen or twenty
species of fossils had been found in the Sutherlandshire
limestones and quartz-rocks, including seven cephalopods,
seven gasteropods, one brachiopod, two annelides, two corals,

and probably a sponge.   Now, of these species it was found
by Mr. Salter that five were certainly and three doubtfully
common to the Lower Silurian rocks of Canada and the
United States, while four might be called representatives of
American forms.

Seeing, then, that the geological horizon of the fossiliferous
rocks could now be so satisfactorily fixed, with what series
elsewhere were the underlying red sandstones and conglomerates to be compared ?   Far down beneath the base of the
fossiliferous Lower Silurian rocks of Wales, in the vast mass

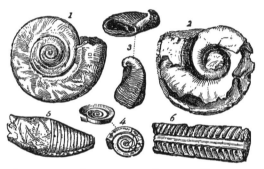

SHELLS FROM THE LIMESTONE OF THE NORTH-WEST HIGHLANDS.

1 and 2. Maclurea Peachii. 3. Operculum of same. 4. Ophileta compacta. 5. Onco-
ceras, sp.   6. Orthoceras (like a Canadian species).

of strata to which the Geological Survey had restricted the
term Cambrian, there lie thick zones of red grits and conglomerates, to which the Sutherlandshire strata might well
be likened.   It is true that in Wales no such break or discordance occurs between the Silurian and Cambrian rocks.
The two series pass insensibly into each other.   But then
in these northern mountains it was apparent from the fossil
evidence that the true base of the Silurian system did not
occur.   A portion of the series, represented elsewhere by
the Lingula Flags, and perhaps by other subdivisions, was

there wanting, so that the marked unconformability between the quartz-rocks and the red sandstones could be satisfactorily explained. Murchison, therefore, proposed to class as Cambrian all the red sandstones and conglomerates lying below the quartz-rocks and limestones of the north-west of Scotland.

But with what could he compare that dark gnarled gneiss which comes out in those strange bare hummocky hills from beneath the vast terraced mountains of red sandstone, so utterly distinct from everything else? Of its vast relative antiquity there could be no doubt. It had been crumpled and crystallized, and then battered by the elements long before even those venerable red sandstones were formed. Hitherto the Cambrian rocks had been the oldest strata of Britain, but now there seemed to be clear evidence of something older still—a yet more ancient foundation-stone underlying and supporting the vast piles of rock which form the mountains of the Highlands. To this lowest rock Murchison gave at first the name of the Fundamental Gneiss. In later years he classed it with the Laurentian Gneiss of Canada—a vast mass of crystalline rocks, which, as far back as 1846, had been recognised by Logan and his colleagues of the Canadian Geological Survey as lying beneath the oldest fossiliferous formations of the colony.[1]

Starting, therefore, from the ancient platform of gneiss found on the western margin of Sutherland and Ross, he traced an ascending section through the undoubtedly Lower Silurian limestones and the vast overlying schists

[1] In 1864 a curious body was obtained from a low part of the Laurentian Gneiss of Canada, and determined by Dawson, Carpenter, and Rupert Jones, to be a foraminifer. No organism of any kind has yet been met with in the fundamental gneiss of Scotland.

SIR WILLIAM E. LOGAN, F.R.S.

and gneiss, up to the Old Red Sandstone of the Highland border. He argued that as this overlying metamorphosed series had Lower Silurian fossils at its base, and was covered by Old Red Sandstone at its top, it could not be anything but Silurian. And thus by one bold dash of the brush, bold, but justified by careful and accurate observation, he wiped out the old conventional mineralogical colouring, which dated from the time when gneiss, mica-schist, and clay-slate were supposed to be necessarily of higher antiquity than any fossiliferous rocks, and substituted for it a mode of representation whereby the great mass of the Scottish Highlands was shown to consist of altered crystalline sandstones, shales, and other strata of Lower Silurian age. No such rapid and extensive change had ever before been made in the geological map of the British Isles.

Working his way eastward across Sutherland and Caithness, Murchison struck the coast again at Langwell, near the Ord. There, under the hospitable auspices of the Speaker and Lady Charlotte Denison, he explored the geology of the Scarabin Hills. Few tracts in Britain suffer more than the treeless moors of Caithness from the biting and furious sea-blasts. In the teeth of one of these storms a party started on horseback, with the geologist at their head, to reach the crest of the only high ground in that desolate moorland. But the weather proved too much for those who had no enthusiasm for quartz-rock. "When Miss S. and the Speaker were driven back," so wrote the leader of the expedition, "I insisted, like an ardent *ci-devant jeune homme*, on continuing to face the storm, until, after climbing along the south face of the middle Scarabin, among the piles of loose quartz-rock, and leading our ponies, unable to hear each other bawl for the roaring of the wind, we went in spite of it to the north side of the third hill, pluck-ing cloud-berries in full ripe fruit. They seem to like the quartz and a full northern blast."

But feats of this kind could not be so lightly performed by a veteran of sixty-six as they had been one-and-thirty years before, when Sedgwick and he scoured these same grey bare hills. The next day the indomitable geologist became a somewhat despondent invalid, chronicling the changes of his pulse instead of the strike and joints of the quartz-rock.

During the leisurely journey southward by Dunrobin and Dingwall to Inverness, and thence by Elgin and Aber-deen into Forfarshire, Murchison's chief employment, besides visiting old friends, consisted in looking after every section of Old Red Sandstone he could meet with. He had resolved to attempt to get that system of rocks in the north of Scot-land into definite order. What measure of success he obtained was shown in the elaborate memoir which he afterwards laid before the Geological Society.[1]

The visits he paid almost always combined a little, some-times a good deal, of geology. Thus when at Rossie Priory, the charming residence of Lord Kinnaird in the Carse of Gowrie, he set out with a party to see the famous sandstone of Dura Den, and compare it with the pale sandstones of Dunnet Head, Hoy, and other scenes of his recent tour in the north. This sandstone has been long celebrated for the abundance of its fossil fishes. In certain layers of the rock, indeed, these relics lie so thickly strewn on each other, and so entire, as to show that the animals must have met a sudden

---

[1] *Quart. Journ. Geol. Soc.*, vol. xv. p. 353.

death, and have been covered up with sand before their scales and bones had time to get loosened and separated. With some labourers to assist, the party set vigorously to work, to quarry out some samples of the ichthyic treasures of the place. Murchison's account of the operation is as follows :—" After finding a few remains and fragments, Lord Kinnaird's eye caught the end of a fish. On quarrying into the rock, after much perseverance the head and a considerable portion of a grand *Holoptychius* came forth, to the exuberant joy of all concerned. The dark and red tints of the scales and bones of this fine large fish shone out in striking contrast to the white and yellow stone. Grand as was this discovery, it was clear that we were only on the threshold, and that by patience the whole fish might be extracted. So thereon we went to lunch under the trees—a most picturesque party, our noble host, having worked harder than the quarrymen, in his shirt-sleeves, and Lady Kinnaird presiding with her attractive manners. We carried off our trophies, hoping still for more, recrossed the Tay, and dined at eight at the Priory—a very joyous party. But our excellent and eloquent explorer of Dura Den [Rev. Dr. Anderson of Newburgh] resolved to complete our gratification. Next morning he went back by rail, and in the evening returned with an immense booty, and all the remainder of the huge fish in a large box completely covered up in wool. This day has passed in uniting the head with the remainder of the animal, cleaning, fixing, cementing, and securing the whole. The animal thus put together measures thirty-three or thirty-four inches by thirteen in width, and is thus considerably larger than the *Holoptychius nobilissimus* which I acquired for the British Museum some years ago."

From such pleasant and restful amusement the transition was abrupt to the bustle and hubbub of the British Association at Leeds. A custom unknown before had now crept into the practice of the Association, viz., that the President of each Section should open the business of the first day by giving an address. Murchison accordingly, having been chosen again to preside over the geographers, had prepared and now read a short discourse on recent progress in geography. From this time onward to the end of his life, in his double capacity of geologist and geographer, he frequently filled at the meetings of the Association the chair either of the Geological or Geographical Section, and furnished on each occasion the now expected introductory address. Indeed, if we group these discourses with those more voluminous essays read each year to the Geographical Society at its anniversary in May, we have material enough to form several thick volumes, consisting entirely of addresses on the general progress of research.

But with such a well-filled set of geological note-books it could not be expected that the President of the Geographers should wholly forsake his old Section C. He gave his brethren of the hammer an outline of his recent doings in the North, more especially with reference to the grand revolution he proposed to make in the hitherto accepted geology of the Highlands. He had brought away in his travelling-bag some bones of the large *Stagonolepis* of Elgin, marked by Agassiz as a fish, but which he shrewdly suspected to be reptilian. In exhibiting these, he renewed the expression of his conviction that in spite of the recognised reptilian grade of *Telerpeton*, and possibly of some of its contemporaries, the strata yielding their remains could not be separated from the Old Red Sandstone.[1]

Though the main question as to the order of the rocks seemed to have been now conclusively settled, strong opposition was offered to Murchison's conclusion by his fellow-traveller, Professor Nicol. A chief part of the reasoning of the former geologist proceeded on the alleged fact that the old gneiss of the west of Sutherlandshire lay deep beneath and totally differed from the later crystalline rocks or so-called gneiss which spread over such wide spaces of the Northern Highlands. At first Mr. Nicol, following in this matter the shrewd observations of Hay Cunningham, had recognised the distinction in question. But he afterwards endeavoured to show that at every point where the "upper gneiss" seemed to overlie the other rocks, it was separated from them by intrusive igneous rocks, and in truth was only the true old gneiss brought up again by great upcast faults. Murchison felt confident that he had made no mistake in his sections. Nevertheless, as he had not seen some localities and sections adduced by Professor Nicol, and as the matter in dispute was one of fundamental importance in British geology as well as in general questions relating to the history of metamorphic rocks, he determined in the summer of 1859 to make a new and final examination of the Assynt and Durness ground, with the advantage of the co-operation of the trained eye of his colleague, Professor Ramsay.

In their northward journey, the steamer, which usually

[1] In spite of the apparent gradation of the Elgin reptiliferous Sandstones into strata of undoubted Old Red Sandstone age, Murchison eventually surrendered this point and accepted them as of Triassic date.

holds her course straight through the Sound of Mull, turned
aside into one of the inlets on the southern shore of that
mountainous island, to land there Maclean of Loch Buy,
who, after a successful life abroad, had come back and re-
purchased the lands of his ancestors, which he was now
about to reoccupy.   In his journal, Murchison refers to this
event thus :—" The gratification I experienced was great in
seeing the happy disembarkation of the rich man from
Java, with his wife, many children, a Malay woman, and all
sorts of traps.   Not to be forgotten that thirty-two years
ago Sedgwick and I danced with our nailed shoes in the
halls of Loch Buy, then belonging to the old laird, who was
ruined.   Property regained by his eldest son.   Guns saluting
—numerous boats."

Much of the ground to be geologized over in Sutherlandshire was the same as that already well explored.   Map in
hand, and tracing out in detail the boundaries of the rocks
in some of the testing sections, the geologists completely
established the truth of the arrangement which Murchison
had adopted, and on the basis of which he proposed to class
the great mass of the crystalline rocks of the Highlands as
altered strata of Lower Silurian age.   The journal of these
rambles, however, wholly devoted to quartz-rock, limestone,
Cambrian sandstone, fundamental gneiss, and other geological matters, affords us no glimpses of its writer in any other
capacity than as an enthusiastic hammerman, up early and
out late, very much pleased and elated to find his main
proposition so completely sustained by his friend and companion's critical examination.

From the mountains of Assynt and the Kyles of Durness
the travellers found their way to the coasts of Caithness and

the Moray Firth, subjecting some of the sections of Old Red
Sandstone there to careful scrutiny, and gradually geologizing
eastward until they landed in Aberdeen in time for the
meeting of the British Association, which was to assemble
there under the presidency of the Prince Consort.

In obedience to a request of the Council, Murchison had
agreed to give one of the evening discourses during the
meeting. He chose for his subject the story of the rocks of

View of the Old Red Sandstone cliffs to the north of the Ord of Caithness.

the Scottish Highlands, and took infinite pains in getting
illustrated diagrams constructed. But popular lecturing was
not his mission. At the close of his address, some interest
was excited by a spectacle unprecedented in the annals of
the Association. A deputation from the Royal Society of
Edinburgh appeared on the platform, and, on the part of the
Society, presented to Murchison the first gold medal, founded
by Sir Thomas Makdougall Brisbane, for the encouragement
of science in Scotland.

Another noteworthy incident occurred in connexion with the Aberdeen meeting. Her Majesty the Queen, desirous of showing her own personal interest in the Association, had invited its members to Balmoral, there to breakfast and witness some Highland games. A goodly number took advantage of the invitation, and duly appeared in the royal demesne, though a succession of drenching showers rather tried their good-humour. Of course Sir Roderick Murchison was not likely to absent himself. The games were in progress. Her Majesty, well wrapt with shawls, kept her place in defiance of the weather, and the Prince Consort moved to and fro, saying a kind word here and there to such of the guests as he recognised. A slight buzz was heard at one part of the grounds, and a knot of eager faces was seen to gather round some central figure. By and bye Sir Roderick emerged, and, obtaining audience of the Queen, announced to her Majesty that he had just received a telegram reporting the discovery by Captain M'Clintock of the Franklin records and log-book !

During the preparation of the memoirs on the geology of the north of Scotland for the Geological Society, and more especially of the little sketch-map of the Highlands which accompanied them, Murchison saw that to complete his work, and show its adaptation and extension to the rest of the country, he ought to compare the clear sections of the Assynt country with others in more southern parts of the Highlands. If he could show that the sequence as made out in the north-west should hold good throughout the rest of the Highland regions, he would not only confirm it but make a great forward step in explaining the structure of a wide and still geologically almost unexplored region,

Accordingly, in the summer of 1860, he determined to put this idea into practice. His original plan was thus sketched in a letter (18th July) to myself, then one of the field-geologists of the Geological Survey in Scotland :—

" Requiring some speedy change of air or absence from over-excitement, I would have liked to have had a real holiday in the Pyrenees or elsewhere. But seeing that Ramsay and Jukes, my generals of division, are both abroad, I have resolved not to quit the British Isles. I propose, therefore, to get away in the middle of next week, and to go to Scotland. Being there, I consider it to be my duty to work out with your assistance the problem of how far the order and classification which are clear and established in Sutherland and Ross are applicable to the more southern parts of the Highlands.

" I think, therefore, of taking you with me in the first instance to Jura and the adjacent mainland, where zones of quartz-rock and limestone abound, and which may prove to be equivalents of my Durness and Assynt Silurians. Having ascertained whether that zone subsides under micaceous flags (as I surmise), I will test the same again between Balahulish and Fort-William. Having settled these points, and having re-explored the heads of Loch Duich, Loch Alsh, etc., I will test the thing again at the head of Loch Maree (one of Nicol's obstructive cases), and, having looked around that tract, will revisit Loch Broom, where the Ross-shire succession is as clear as that of Sutherland. Finally, we will cross to the Lewis, where I wish to satisfy myself still more conclusively as to the fundamental gneiss."

This programme was adhered to in the main, but considerably extended. By a series of traverses across the Highlands, an attempt was for the first time made to show the general geological structure of that region, starting from the old gneiss, on the north-west side, and passing up to the overlying Old Red Sandstone on the south-east. The full details of this campaign were afterwards worked out in a long conjoint memoir, read to the Geological Society.[1]

The veteran geologist was now in his sixty-ninth year— rather an advanced age for a return to the rough fare, and still rougher ground and weather, to be looked for in the remoter parts of the Highlands. But his enthusiasm remained with seemingly no abatement; his eye still kept its wonderful quickness in detecting the really salient features of the geological structure of a district; his powers of walking, though now of course manifestly on the wane, were still equal to the accomplishment of a ten or twelve mile tramp, while his general strength and capacity for endurance continued to show the singular vigour of constitution to which through life he had owed so much. This, however, was his last great expedition, and he felt at the time that it must be so. He knew of nothing else likely to tempt him into the hardships of the field again, and so at times he put on an extra stimulus to carry him bravely through fatigue and discomfort, which bade fair to land him at last at his goal— the orderly grouping of the rocks of his own Highlands.

Of personal incident the tour furnished little. Yet it brought out in strong relief, and enabled me to realize various features of Murchison's character, and to gather traits and anecdotes which have been already woven into this narrative. It laid the foundation of a sincere and warm friendship between us. He had previously known me only

[1] On the Altered Rocks of the Western Islands of Scotland and the north-western and central Highlands.—*Quart. Journ. Geol. Soc.*, 1861.

as one of his younger officers, who was in the habit of spending his holidays in geological exploration, and communicating the results to the Geological Society. From this time forward he treated me with almost paternal kindness, frankly taking me into his intimate confidence, and showing on many occasions a thoughtful and tender solicitude for my welfare, which has endeared the memory of his friendship as one of the brightest recollections of my life.

His own journal of the tour, like the rest of his journals in this latter part of his life, contains scarcely an entry save what is geological. Its pages, after the interval of years, recall to me the eagerness with which he pursued his quest, the shrewdness with which he could guess at the probable structure of a hill several miles away, where most eyes would have detected nothing, but where, after a good hard climb, one found his conjecture to be true; the pertinacity with which, in spite of the attractions of Highland sport and Highland hospitality, to both of which he yielded so far, he yet held on his way until he had accomplished his task.

A few of the incidents of the tour which impressed themselves on my memory, though of little note, may perhaps fitly find a place here. No one who has often heard Sir Roderick Murchison address public meetings can have failed to notice how characteristically his Highland blood would show itself. He was proud of being a Highlander, and seldom lost a chance of proclaiming his nationality. Back in that picturesque region of Kintail and Lochalsh, where his forefathers had lived, his patriotism glowed again with renewed ardour. He revisited, with ·undiminished interest and pride, the scenes where Donald Murchison had baffled the King's troops. When we were

together in Loch Duich, though no geological necessity
called him, he must needs once more make a pilgrimage to
the Bealloch of Kintail.   Standing on a rising knoll, his
left hand stuck into the arm-hole of his waistcoat, and his
right holding a stout staff, with which he pointed out the
leading features of interest, historical or geological, his face
would kindle with the old martial fire as he went over again
the events of "the '15." In the same spirit, he solicited and
obtained from the proprietor of the ground leave to choose a
site for a monument to commemorate the deeds of his illustrious kinsman.   A bright autumnal afternoon was given to
that pious quest.   We went by boat, creeping in and out
among the islets and promontories at the mouth of that
wonderfully fine inlet, Loch Duich, and fixing at last upon a
knoll of rock amid the heather and bracken, from which we
could look over to Eilan Donan Castle, and away up to the
mountains of Kintail and Glenelg on the one side, and over
to the peaks and glens of Skye on the other—a site which
the annual crowds of steamer-carried tourists would be sure
to see when the obelisk should be placed upon it.

Again—the Murchison sept had not been all as prosperous in the world as the Laird of Tarradale.   Some of them
still remained in the original district in humble circumstances, but, with the genealogical skill of true Highlanders,
they could yet count their kinship to the geologist.   I
remember, on one of our excursions, we halted at some fern-
thatched cabins, and were met by one or two plaided cottars,
with whom I left Murchison in talk.   I was told afterwards
that they were some of his distant relatives, or clansmen,
whom he always visited and assisted when he returned to
that part of the country.

While still in this Ross-shire district, we attended the
English service in the parish church, which, in spite of a
very rainy day, was crowded by a Gaelic congregation of
some 500 people in wet clothes, who gave a good illustration of that loud and deep groaning during the sermòn,
which is sometimes so marked in the churches of the north-
west Highlands.   Murchison's face was a curious study
during the service.   Naturally reverent, and evidently with a
strong desire to compose himself to the frame of mind and
posture of body proper to the occasion, he yet wore a
droll expression of wonder as he watched the gravity of the
hearers amid sounds which, anywhere else, he would have
supposed indicative of the deepest anguish or pain.   When
the service was ended, and we were again in the pure air
outside, he drew a long breath, and remarked to me that
it was the last time he should ever enter a Presbyterian
church !

Next Sunday was as fine as the previous one had been
wet, and was very differently spent.   We had reached one of
the wildest tracts of Western Ross-shire—the mountainous
and broken country between Lochs Torridon and Maree—
interesting in its singular scenery, and specially interesting
in its clear, as well as complicated, geological sections.   It
was one of the districts which Murchison had particularly
marked as likely to test his views of the order of succession
among the rocks of the north-west Highlands.   The previous
day, unable to obtain any kind of conveyance when we landed
from the small boat, which had brought us to the upper end
of Loch Torridon, we had crossed on foot from that fjord to
the little inn of Kinlochewe, at the head of Loch Maree—
a tramp of twelve miles, which the veteran accomplished

easily. Indeed, so much of geological interest lay crowded
round us by the way, that the gleaming waters of Maree
seemed to come almost too soon in sight. Next morning
ushered in one of those days of which perhaps not more
than half a dozen fall within the life and experience of any
ordinary man.   Under a sky of deep clear blue, every peak
and crag even of the far mountains stood up sharply marked
by its light and shadow.   Not a branch of the grand old
pines, not a spray of the deep heather and autumn-tinted
bracken, stirred in the quiet air. Loch Maree, stretching
away for miles down the valley, lay without a ripple, save
what were made now and then by the active water-fowl,
and reflected without distortion every one of the many
varieties of form and colour which diversify the sides of its
grand circuit of mountains.   In the midst of this general
calm, and as if in some vague way protesting against it, the
splintered crags and ruin-covered slopes, rose all around.   It
was such a morning as calls up most strongly the desire for
solitude, and moves most deeply such sympathy as may be
in a man for the mingled beauty and grandeur, tenderness,
and power, by which nature appeals to what is best and
noblest within him.

We breakfasted, and at once separated. Sir Roderick
spent the day sauntering by the head of the Loch along the
base of the mountains on its eastern side, and sitting down
now and then to take a rough sketch of the strange and
picturesque mountain-outlines of Ben Eay and the western
side of the valley. For myself, I longed to be on the summit
of one of the far-gleaming peaks, and set off accordingly.
The fresh buoyant air of the mountains; the huge masses of
red sandstone capped with white quartz, like sheets of peren-

nial snow ; the depth of the glens, once filled with glaciers, and still strewn with their moraines ; the ruggedness and dislocation of the slopes and cliffs ; the solitude of the whole scene, broken now and then by the bound of a group of red deer startled from a favourite corrie, or by the whirr of the snowy ptarmigan ; the ever-widening panorama of mountain-summit, gorge, glen, and lake, as each higher peak was gained in succession ; and then, from the highest summit of all, the vista of the blue Atlantic, with the faint, far hills of the outer Hebrides, and the nearer and darker spires of Skye,—all this, added to the absorbing interest of the geo-logy, filled up a day to the brim with that deep pleasure which becomes a life-long possession. Night came down when the inn at Kinlochewe still lay a good many miles on the further side of a tract of mingled mountain, glen, river, and bog, through which lay no road. Fortunately, in the end, the moon rose, and the inn was reached somewhere near midnight.

The delay in the return of his companion gave Murchi-son not a little uneasiness. As hour after hour passed, he grew so impatient that he began to insist on some of the people at the inn turning out with lanterns. His remon-strances, however, met with a sullen indifference, very unlike the usual attentiveness of the household. It turned out in the end that the want of sympathy sprang from a theological cause : " It was the Sabbath-day ; the gentleman shouldn't have gone to walk on the Lord's day." In short, the gentleman, had he been lost, would have deserved his fate, and would have furnished to the pulpits of the district a new and pregnant illustration of the danger of Sabbath-breaking !

From these north-western tracts, where the chief geological task was to run the boundary lines, which had already been well-defined in Assynt, down through the mountains of western Ross-shire to the Kyles of Skye, we struck southwards and made a series of traverses across the central Highlands. The object of these further explorations was to discover, if possible, the structure of the great Highland region, by using the key which the north-western sections furnished. Nothing more was aimed at or could be attempted than the ascertaining of the general fact, that the metamorphic rocks of the Highlands, enormously thick as they appear to be, have been so crumpled and folded that the same zones reappear again and again in successive great arches and troughs. And this was successfully accomplished. Instead of a chaos of dislocation and crystallization, these Highland rocks really proved to have, on the whole, a very orderly arrangement, and could be recognised and traced, group by group, for wide spaces across the country. This piece of work formed the natural complement to what had already been effected in Sutherlandshire. It bore pregnantly, too, not merely on local geological questions, but on that wider and difficult question, the history of metamorphism. The rocks of more than half a kingdom, which had previously been mapped and described merely as so many mineralogical masses, covering certain spaces of country, and belonging to the so-called "primary series," were now found to be really the altered equivalents of ordinary sedimentary strata, retaining still abundant evidence of their origin, and, despite their severe alteration, capable of being traced and mapped in the same way as rocks which had not undergone any such change.

GENERALIZED SECTION ACROSS THE NORTH OF SCOTLAND.

(*Siluria*, 4th Edit., p. 169.)

SUTHERLAND AND ROSS-SHIRE

CAITHNESS.

Loch Inver. Suilven.   Assynt.   Ben More.   The Ord.   Caithness.   Hoy Head.

a, b c^1 c^2 c^3 * c^3 d^1 d^2 d^3

Laurentian.   Cambrian.   Lower Silurian.   Lower Silurian.   Granite.   Old Red Sandstone.

a Fundamental or Laurentian Gneiss; b Red and Purple Cambrian conglomerate and sandstone, lying on the eroded edges of a, and covered unconformably by c^1, the lower quartz rock; c^2, the fossiliferous limestone; c^3, the upper quartzose, schistose, and gneissose rocks, all of Lower Silurian age. These strata are pierced here and there by granite *, as at the Ord of Caithness. Towards the east they are overlaid by the Old Red Sandstone, consisting of d^1, lower conglomerates and sandstones, d^2, Caithness dark grey flagstones, d and d^3, upper red and pale grey or yellow sandstones, well seen at Duncansby and Dunnet Heads, and in the cliffs of Hoy, one of the Orkney Islands. [Murchison believed that he had traced a passage from d^2 into d^3. Researches in those northern regions in the autumn of 1874 by the writer of this biography, accompanied by his colleague, Mr. B. N. Peach, have proved that in reality there is a complete discordance between the Caithness flags and the overlying Upper Old Red Sandstone.]

Nevertheless, though the general law of structure in the Highlands came out with great precision as the result of this autumn's work, the difficulties of detail were and could not fail to be very great, and in many cases, with only a limited time to wrestle with them, insuperable. Murchison, full of the grand truth which our sections had made evident, refused to deal with any of these, or contented himself with a general invocation of gigantic fractures and reversals. It was vain to point out that any number of such disturbances would not solve the problem, and that probably in the south-eastern Highlands the mineral character of the typical zones of Assynt is replaced by something different, or at least that higher zones come in there which do not appear to the north-west. He stuck to his leading principle, from which no amount of contradictory detail would make him swerve. And, so far as regarded his object at the time, he was doubtless right. I find among his notes references to some of these attempts on my part to persuade him that everything did not fit so harmoniously into his views as he wished and imagined. Thus :—" Doubts and difficulties of Geikie. I see nothing but what I expected in approaching the Grampian *jam.*" Again, in the Blair Athole district, where, with our limited time, and the puzzling character of the sections, I was driven to despair, he records,— " Poor Geikie is cracking his brains and exhausting his energies in trying to coax these frightfully chaotic assemblages into the order of the north-west. Some day he may accomplish it. We shall see what he says of the chief limestone of Cairnwell, and its relations to quartz-rock."

But in spite of these unsolved difficulties which still await the resolute and patient toil of a geologist furnished with accurate topographical maps, the tour had been abundantly successful in its leading object. The general structure of the Highlands could no longer be considered doubtful. A detailed narrative of the observations was prepared, in the form of a conjoint memoir to the Geological Society. As a further result of this tour, the plan was conceived and soon put into execution, of preparing a small geological map of Scotland, to embody, in a broad and generalized form, the new views of Highland geology. This little sketch-map was the first, and as yet the only one, in which the rocks of the country from bottom to top are treated in rigid stratigraphical order, and delineated so as to show the structure of the country.

With this Highland tour, and the preparation of the narrative of it for the memoir, Murchison closed the last great geological task of his life. It was a worthy end to so long and active a career. Apart from its importance in a scientific point of view, there was a fitness in the fact, that after wandering far and wide, and gaining distinction from the furthest confines of Europe, he should return to his native Highlands and gather his last laurel from the rocks on which he was born. Probably no aspect of the matter gave to himself more pleasure than this.

CHAPTER XXV.

OFFICIAL LIFE IN LONDON.

WHEN the main interest of a man's life has lain in its external incident, and when chiefly through that outward work and movement have his own character and development been shown, there must needs come a time when, as the life grows less and less eventful, it presents fewer and fewer features for a biographical record. The man may have had no ever-advancing inner life; or, at least, may have kept it so shrouded within his own soul that even his most intimate friends could not trace its workings and progress. He has lived, and moved, and had his being in the stir and bustle of the world. He has had perhaps a great work to do, and has done it. The ardour and enthusiasm which formerly braced him for effort, and brought freely into play the varied feelings and faculties which went to make up his character, have now mellowed down. Such a one no longer talks with hope and exultation of what he is going to do, but rather dwells with complacency and pride on what he has done. There may still be, perhaps, a great interest about him, but it is mainly retrospective. The circumstances probably no longer exist around him to call out those qualities in him which made him different from his fellows, even if his energies were equal to renewed exertion. And hence, perhaps, in ordinary general company, or even in more quiet and familiar converse, an observer may fail to detect those qualities, may even rather fix upon others of a less pleasing kind, which, once overshadowed by the nobler growth that made the man what he was, have since sprung into undesir‑ able prominence.

We have now arrived at this comparatively uneventful period in Murchison's career. Not that he grew less busy or less devoted to the pursuits to which he had given so many years of energy and enthusiasm. But he had achieved the great work of his life. His remaining years were per‑ haps even busier, as they certainly were fuller of distrac‑ tion and multitudinous cares than any of those which preceded them. But there no longer ran through them the connecting thread of one central idea to be worked out, one dominant task to be performed. He gave himself up to the calls of his official position in London; and in accepting its dignity and honours he had to undertake also its numerous and often burdensome duties.

In the later years of his life, therefore, we see him in a capacity different from that in which he appeared during any of the previous periods. He had long held a prominent place in London society, but he never regarded the due maintenance of that position as the chief function he had to perform. It ranked, in his estimation, after his scientific work. Year after year he had fled from London life to renew his labours in the field among his beloved rocks. But the claims of the metropolis had been fixing their hold more and more firmly upon him. His annual escape to the country

had been each year growing later, and his return to town
getting earlier. The duties of the Jermyn Street establishment kept him indeed in London during most of the summer.
In his last ten or fifteen years, therefore, his work lay mainly
in London,—very miscellaneous work, and hardly capable of
any methodical record in these pages, but yet in many ways
helpful to the progress of science and the recognition of its
claims.

It is not however merely, or even chiefly, the want of
incident which makes this concluding portion of Murchison's
life more difficult of satisfactory biographical treatment.
There is an almost total want in it of material of a personal
kind. He kept no diary. His letters, though probably
more numerous than ever, treated almost wholly of scientific questions or matters of business, and give us but a feeble
glimpse at the man himself. We are thus thrown back upon
his public career, wherein the same qualities remained conspicuous which had marked the previous and more strictly
scientific period of his life. As one year passed very much
like another, we can no longer conveniently follow a chronological order, but must be content to group, under separate
heads, the various kinds of activity by which this concluding
portion of the veteran's life was distinguished.

Probably the most effective grouping will be into four
sections :—1*st*, Official work, either in connexion with the
Geological Survey, or arising out of Murchison's position as
a geologist. 2*d*, Holiday-rambles, when he escaped from
London to the country or to the Continent. 3*d*, The chair
of the Geographical Society, and the efforts he made, when
holding it, to further the progress of Geography, and aid the
expeditions of travellers. 4*th*, The last touches to his geo-

logical work at home and abroad. The present chapter will be devoted to the first of these sections.

So far as the Geological Survey was concerned, the duties of the Director-General were twofold. He had to conduct the official correspondence and other office-work at Jermyn Street, and he was expected, as occasion might require, to visit his officers in the field, and confer with them in any matter requiring special consideration. Murchison did not carry into his task the undivided earnestness and almost enthusiastic energy of De la Beche, the founder of the Survey. He joined the service at too advanced a period of life to make this possible. Consequently, though he did as much at Jermyn Street as could have been reasonably expected of him, he rather avoided the more exacting claims of the field-work.

One question to which he devoted a good deal of time and thought was the increase of the force under his command, with a view to greater expedition in the surveying of the country. With the advance of geological knowledge there had been a corresponding progress in the style of mapping, especially in the direction of far greater elaboration and detail. In former years geologists, who set to work to make a geological map, usually omitted notice of the superficial detritus, and contented themselves with tracing, as well as they could, the boundaries of the underlying rocks. But now the previously neglected surface deposits received a yearly increasing share of attention; while, at the same time, greater perfection and precision appeared in the representation of the more solid formations lying underneath. Hence geological surveying became a more and more laborious occupation, demanding an increasing measure of time, toil, and skill.   With this tendency, it was evident that the Geological Survey could not be expected to show much increase in its rate of progress.   On the contrary, that rate must necessarily diminish, unless the working staff were increased.   To obtain from Parliament such an augmentation of force as would materially accelerate the progress was the task to which the Director-General now addressed himself.   In Scotland there had been for several years a force of only two surveyors.   So small a staff had been necessary at first, for the maps of the Ordnance Survey, on which the work of the Geological Survey is traced, were not ready.   But this difficulty no longer existed to the same extent; maps could now be had for a great part of the central counties, and every year largely added to their number.

Reflecting on these matters, and, like other heads of departments, having a wholesome fear of sharp inquiry on the part of the House of Commons, the Director-General, in 1858, propounded his views to Professor Ramsay, his second in command :—" Seeing and believing that the time is fast approaching, if not actually come, when Scotland must be put on the same footing (*geologicè*) as Ireland—or I shall never be forgiven by my countrymen,—and knowing that we ought absolutely to have at least the whole of our present force of field-surveyors to do justice to England and her large unexplored and most valuable regions, I have quite made up my mind to ask for a considerable increase of our forces.

" As I am thoroughly convinced that a very great addition to the staff both of England and Scotland is desirable, so it will be my duty to recommend such in my ensuing annual report. If this should not be granted, the Survey will not be finished for fifty years. . . . I am continually upbraided with not bringing out the [map of the] metropolitan districts, and the publication of the Edinburgh sheet before that of London will add to the discontent. It is quite certain that the Scotch members will clamour for more work, and if I reply that we have two men only for the ' Land of Cakes,' they will insist on justice equal to that afforded to Erin. . . . Being responsible for the effecting of more in my time (seeing the vastly greater amount of [published Ordnance] maps) than could be done before, I cannot sit still on the old status."

As the result of his efforts he succeeded in getting a considerable increase of force at this time. The progress of the Survey, accordingly, continued to improve. But though the number of assistants was augmented, the annual amount of country surveyed had not increased in the same proportion. This arose partly from the much greater detail and perfection which the method of surveying had reached, whereby, of course, greater time was needed than before for the same area of country, and partly from the inducements to the officers of the Survey to quit their hard work and small pay for other and more remunerative appointments, so that the service was deprived of efficient surveyors, and necessarily lost some time in training their successors. In the latter part of the year 1866, the Government of Lord Derby took into consideration the whole question of the condition and progress of the Geological Survey, and, in concert with the Director-General, prepared a scheme for its re-organization and enlargement. By the new arrangement, the staff in Great Britain which had hitherto been under the charge of Professor Ramsay, as

Local Director, was now divided into two parts, the larger
of which, for the survey of England and Wales, remained
with him as its Director, while the remainder, for the survey
of Scotland, was placed under the supervision of a new
Director,[1] the Irish branch remaining under its former
Director, as before.   Each of the three branches received a
great augmentation to its staff, so that the total force of the
Survey in the United Kingdom was raised from 37 to 75.

In the narrative of the history of the Geological Survey
given in the twenty-second chapter, it has been shown that this
branch of the public service had been several times transferred from one Government department to another before
Sir Roderick Murchison was put at its head.   Shortly after
his accession to office still another change was meditated.   At
that time, as part of his office-work, he took great pains to com-
bat a proposed transfer of the Geological Survey and Museum
of Practical Geology from the control of the Board of Trade,
under which they had now thriven for some years, to the
Education Department of the Privy Council.   Theoretically
there might be no objection to the change, but he saw, or
thought he saw, the advent of a time when the scientific
character of the institutions under his charge would be
dealt with by men who had no knowledge of or sympathy
with science, and whose control would fetter the natural
and free development alike of the Survey and of the School.
He drew up a lengthy document, in the form of a letter to
the President of the Board of Trade, in which, after setting
forth the history of the Jermyn Street establishment and the

---

[1] Having for a number of years previously had the chief conduct of
the Survey in Scotland under Professor Ramsay, I was named to the new
post by the Director-General, and appointed in April 1867.

good work which it had done and might yet do, he states his fears in the following official form :—

" Liberal as the Minister may be under whose control the general education of the nation may be placed, there is little doubt that in this country the greater number of its instructors will be drawn from among such of the graduates of the ancient Universities as, both by their training and position, must be to a great extent disqualified from assigning their due importance to the practical branches of science. Such persons may be eminent in scholarship and abstract science and yet ignorant of the fact that the continued prosperity of their country absolutely depends upon the diffusion of scientific knowledge among its masses. They may, with the most sincere and earnest intention, not only fail to advance, but even exercise a retarding influence on such diffusion, and may object to a course of study which, as now pursued, is irrespective of religious teaching. Experience has shown in how sickly a manner practical science is allowed to raise its head under the direction of those persons whose pursuits are alien to it, whilst in every land where it has had due support the greatest benefits have resulted.

" Placed as the Geological Survey and its affiliated branches now are, in subordination to the Board of Trade, they are continually aiding in the development of an amount of mineral wealth far exceeding that of any other country, and in this wholesome and important action the movements of our body are not only unfettered, but are likely to receive all that encouragement which seems alone to be wanted to enable this establishment to be eminently useful in instructing that class of persons who will materially augment the productive industry and trade of Great Britain."

This letter, Murchison says, was lost in one or other of the Departments into whose hands it came, so that when it was required to be printed, in obedience to an order of the House of Commons, a copy of it had to be obtained from the letter-book of the Survey Office.   Such treatment of it did not indicate that it had had much weight, so that he could hardly be surprised shortly afterwards to find himself and all his establishment transferred bodily to the custody of the Science and Art Department of the Privy Council.   He continued up to the last to lament the change.   It led, he thought, to one of the very evils he had predicted, inasmuch as it placed him and all his Professors practically under the supervision of men who had no knowledge of, and probably as little interest in, scientific progress.   But the change was, after all, more apparent than real, and probably his strong objection to it was in good measure personal.   Previously the Director-General of the Geological Survey had reported direct to a Minister of State, now he would have to conduct his communications through Mr. Henry Cole.

Over and above the ordinary and daily routine belonging to such an office, the position of Director-General of the Geological Survey necessarily involved an accession of those incidental duties and interruptions which every man in a public position must expect, and which, as they often consume much valuable time, demand the exercise of no small portion of good temper.   Without attending to the chronological order of the incidents, let us gather from his letters as characteristic a picture as may be obtainable of the heterogeneous nature of these various occupations.

Frequent communications of an official kind passed between the Foreign and Colonial Offices and the Jermyn

Street establishment relative to mining and cognate matters abroad or in the Colonies. Now and then, indeed, the correspondence even became a friendly one between Murchison and the Colonial Governors or other authorities. Thus with Sir Henry Barkly he kept up an active correspondence regarding the gold-diggings in Victoria, during which he saw reason to modify, or almost to abandon, his dogma that deep mining for gold in the solid rock can never be pursued to profit.

To the same correspondent he writes on Colonial defence :—

" I am glad you approve of my resounding the tocsin as to the defenceless state of your noble Colonies. The old mother is so apt to fall asleep, and fancy that nothing can ever befall her, that it has been a hard matter to rouse her to call all her own sons at home to arm in her defence. Now perhaps we are running into the other extreme ; and though I am delighted to see the martial spirit called up in the Volunteer movement, I confess, as one who disembarked in Portugal in 1808 alongside of Sir A. Wellesley, and who has thought much upon his old profession of arms, I have little reliance in any irregular, desultory, unconnected, and ill-disciplined aids. I adhere to the doctrine,—' Dieu est toujours avec les gros bataillons,' and I heartily wish the Government had turned out 40,000 or 50,000 good militia-men."

With Sir William Denison also, the Governor-General of Australia, he exchanged occasional letters on Colonial geology, and other subjects. Thus to the latter correspondent he writes :—" Your letter stimulating me to exertion in favour of the publication, by the Government, of the natural

history of the British Colonies came unluckily just as our
Ministers were in an agony about their untoward Reform
Bill, and since then Sir Edward has been so unwell that I
have as yet been unable to aid you.   When we look at the
splendid publications of the Yankees respecting the geology
and natural history of their several governments, it is humi-
liating to be forced to confess that Britain does so little in
this line.   I confess that I see little prospect of inducing our
Government to undertake such a scheme for all our colonies,
though I have hopes that the colonies which specially
called for and paid their geologists and naturalists would be
assisted in any publication by the Imperial Government. . . .

" Just as Trinidad paid half the salary, and the Imperial
Government the other half to the surveyors, so the Colonial
Government Office might be at half the expense of the publications.   Now this is a practical measure as relates to one
colony at a time, and in one region of the world ; but when
you talk of these geologists coming under my control as
principal editor, you have no idea of the labour that this
would entail."

" . . . You will perceive that I appeal in favour of
a more efficient maritime protection of our long and exposed
sea-board.   It is an old hobby of mine, and I cannot yet
divest myself of the apprehension that we have been too
heedless of the increase of power of our Gallican allies in
waters where they have neither colonies nor commerce.
I had a note about the Fiji Islands, but erased it in consequence of some talk with one of the officers of the Austrian
frigate ' Novara.'

" This is my valedictory Geographical Address, the preparation of which involves too much labour on the part of

an old geologist, who has many other duties upon him, both scientific and official.

" The last three years of my life (as regards my own career as a geologist) have been chiefly spent in preparing and working out a new classification of the rocks of my native country, the Highlands of Scotland, and I am now endeavouring to give the last touches to this labour of love and hard work preparatory to the meeting of the British Association at Aberdeen, at which I shall have to hold forth on the subject."

But Murchison, though an indefatigable correspondent with his friends and acquaintances, had lived too long apart from the forms and routine of official business to get very readily into the ways and style usual in public departments. His letters were apt sometimes to look like embryo lectures, or even like bits of some of his presidential addresses. It was not encouraging to him consequently to receive three lines of official acknowledgment in answer to a document which perhaps covered a good many foolscap pages. Now and then too, being on a personal and friendly footing with the heads of other public departments, he would write to them on public business, but in such a way as to leave them in doubt whether he meant his communication to have an official character. Of all those with whom, by virtue of his position in the Survey, he had to come into frequent official relations, there was none who used such plain language to him and of him as the Comptroller of the Stationery Office, the late J. R. M'Culloch. That stern guardian of the public purse had no sympathy, or even patience, for the Survey's scientific publications, which, in obedience to Treasury orders, the Stationery Office had to

issue to the public. The MS. of some geological memoir,
which had exercised perhaps the collective wisdom of the
Survey, and had just been stamped with the " imprimatur" of
the Director-General, would be received by him with a gruff
—" Well, some more of Sir Roderick's —— trash ! " The
same caustic critic could be as sharp with his pen as with
his tongue. Witness the following plain-spoken but no
doubt well-timed and obviously sensible note :—

" MY DEAR SIR RODERICK,—I received your letter,
marked ' Confidential,' and I have done what I thought
was right under the circumstances. Confidential letters
are very awkward things in matters of business. The person
to whom they are addressed can't, and the person by whom
they are addressed won't, act upon them. Hence they had
much better be withheld.—I am most truly yours,

" J. R. M'CULLOCH."

The affairs of the Geographical Society occupied during
all this time a chief share of Murchison's time and thought.
Their interest and importance, however, demands separate
treatment, and they will therefore be more particularly referred to in a later chapter. It may be noticed in passing
that besides the more distinctly geographical tasks or routine
duties of the President of that Society, Murchison took
occasion still, as we have seen he used to do, to mingle as
much sociality and good-fellowship with the proceedings as
he found it possible to introduce. And the experience gained
in this way he would now and then offer to a friend who
had the same sort of task to perform. Nobody in London
had had more experience than he in presiding over meetings,
whether dull and scientific, or lively and social, so his advice

was a useful guidance. The great Livingstone Festival of 1858 was a good example of his happy tact in this kind of duty. Livingstone was about to start on a new mission of discovery, and it had been determined to give him a more thorough outfit. Murchison took an active interest in the preparations, and assisted in procuring a young assistant to accompany the intrepid explorer as geologist.[1] Escaping from these cares to the quiet of the country, he writes to Professor Phillips just before the anniversary meeting of the Geological Society :—" 17*th February* 1858.—I came here [Chertsey] to get my gullet into order, the severe changes of weather, and my great exertions in getting up and carrying through the Livingstone Festival, having done me up. I am now called up again to settle some dispute about the salary of one of L.'s followers.

" Your note of 15th, just received, augurs well, and promises to make your accession more glorious than that of any of your precursors.[2] With three such public men as you have secured, you need give me little to do.

" One piece of advice I seriously give you. There is nothing so fatal to a public dinner (*crede experto*) as a plenitude of toasts. Ten should be the outside, including the Royal and Loyal. This was my number at the Livingstone Festival, and by my precision of firing, *i.e.* never losing time and yet giving them time to breathe, I got through before or just at midnight.

[1] The assistant eventually chosen was Mr. Richard Thornton, a student of the Royal School of Mines, who afterwards died in Africa while serving under the Baron von der Decken.

[2] Professor Phillips had been chosen President of the Geological Society. The preparations here mentioned refer to the anniversary dinner of the Society, at which he was to preside. Murchison's advice, and his allusion to his own methods of procedure, are very characteristic of him.

" I would give Cardwell a toast to propose, and not let
him reply. The toast he could best give (rely upon it) is
' The Geological Survey and the Government School of Mines.'
Having been himself the Minister under whom the whole
concern acted, it is just the subject he will like to speak upon,
and if you do this, and have a reporter at the dinner, you
will do us in Jermyn Street, *i.e.* your old shop, real service.
As you are taxed enough, I will send a letter to the editor of
the *Times* with a passport in my name for a reporter of
the great Leviathan."

In the preparations for the International Exhibition of
1862, Murchison, from his official position at Jermyn Street,
and in connexion with the department of Science and Art,
necessarily had his share.   He was chosen chairman of one
of the juries, and in that capacity had his hands for a time
kept pretty full of work.   To his friend, Sir Henry Barkly,
he writes :—

"The Commissioners of the International Exhibition of
1862 have applied to me to know if it will not be possible
to test in this [the Jermyn Street] establishment the economic value of the various coals of the British Colonies, of
which, in the event, the Governors would send specimens.
This is indeed an important affair, and I will endeavour to
have it carried out, provided the Home Government will pay
for the cost of the inquiry.   We have no staff in this building, nor any space for such an inquiry (on the great scale).
We must, in fact, have ground, set up large boilers, and
employ several chemists, etc. Nor can it be done in a hurry,
if a really valuable result is wanted. . . . I find that our
metallurgical Professors think we can sufficiently analyse the
various Colonial coals for general purposes of comparison

without going into the tedious and expensive details formerly employed by Sir H. de la Beche for our own navy." . . .

" Geological surveys are all the fashion in New Zealand. I have already sent out Dr. Hector to Otago on a three years' survey, with a good assistant, and I have no doubt he will do capital work. His portion of the labour in defining the character of the upper region of the Saskatchewan and the Rocky Mountains, also of British Columbia, was admirably done.

" I have now an application from Wellington province for another surveyor. In replying thereto, and hunting out a fit man, I could not avoid the expression of my satisfaction in reading two reports in the *New Zealand Gazette*, by the Honourable L. C. Crawford, on the geological structure of the province of Wellington."

" 23*d June* 1862.—Here I am again President of the Geographers, my eighth year of office. I presided over 200 people at dinner in Willis s Ball-room, and as I had some of the foreign chairmen of classes in the International Exhibition, I contrived to make the evening pass with liveliness and point. Gladstone spoke admirably, but they scarcely noticed his speech, and omitted all my sayings and doings in the *Times.*

" I have had, besides, very hard work as chairman of Class I. of the Exhibition. With 2000 exhibitors in my class, it has been no small difficulty to adjudicate with fairness some 300 or 400 medals. The weather has been positively horrible,—wet and cold rains for ever, but to-day there seems to be a genial change. It is marvellous that the Exhibition should succeed as well as it does, despite the Palace shut up, the Court absent, and half of our cotton-mills closed." . . .

Again, when the outcry arose regarding the probable early exhaustion of coal in Britain, and a Royal Commission was appointed to investigate the subject, Murchison was nominated a member of the Commission, and acted as chairman of two of the committees into which the Commissioners subdivided their number for the purposes of the inquiry. In the report finally adopted and printed, the opinion was expressed that a productive coal-field probably exists under the Chalk and other Secondary rocks of the south-east of England. Against this statement Murchison strongly protested, and his protest was appended to the report. He believed that though Carboniferous strata might be found at no great depth, geological analogy was wholly adverse to the idea of any productive coal-basin ever being found in that part of the country. We shall probably have this question settled at no distant date, when the present Wealden boring shall be completed.

As one of the Trustees of the British Museum, Murchison took a keen and active interest in the management of that great institution. Thus he writes to Sir Philip Egerton :— " I regret much to say that the electors of the Trustees of the British Museum gave us yesterday not only L—, but also W—! as if it were nothing but a receptacle for Whigs and Tories! I had written urgently to the Archbishop and the Speaker, and had recommended Lyell and Darwin, or either of them. I hear that the Archbishop proposed them both, but was beaten."

By way of illustration of some of the minor but often useful and friendly occupations for which the Director-General of the Geological Survey found time, reference may be made in conclusion to the trouble he seemed to enjoy in promoting a subscription-list or testimonial for a poor fellow-worker in science. Thus, in the winter of 1858-9, Auguste Balmat, the prince of Alpine guides, under whose tuition it will be remembered that Murchison had done his share of glacier work, came to London, and among those who had profited by his thoughtful and sagacious care the desire arose to present him with some mark of their gratitude and esteem.[1] A small committee, with Murchison as one of its members, was formed to carry out this design. To the Master of Trinity Murchison writes about the matter thus :—" I specially introduce the name of James Forbes on the committee as the leading man. He indeed it was who made Balmat what he is, and most sincerely is Balmat attached to Forbes. I specially wish to consult him about the testimonial, and I know it will gratify Balmat that his old master's taste has been displayed in the matter."

To Forbes, on the same friendly mission, he says :— " A. Balmat has just arrived. He called on me yesterday, and I found that the present which would most gratify him would be a photographic apparatus, which he would employ in delineating some of the striking physical features of the Alpine regions."

Besides such miscellaneous occupations, Murchison did not forget his old and favourite science. But before we trace the record of his last touches to Geology, we may turn in the next chapter to see how the holidays from his official work in London were now spent.

[1] For an interesting account of this skilful guide and most excellent man, see a letter by Mr. Alfred Wills, Appendix C to *Life of J. D. Forbes.*

CHAPTER XXVI.

AFTER his accession to the office of Director-General of the Geological Survey, Murchison never took any of the prolonged continental journeys which had marked the previous periods of his life. There was still much to be done among his own palæozoic formations abroad, much which he might yet accomplish himself. But he now found it by no means so easy to get away as it used to be. He succeeded indeed in escaping from London for a month or more every year, but he could not often go to France or Germany. The attractions of the Highland geology, as already narrated, carried him north for several years, and he there worked out his last important piece of field-geology. But the Highland labours were not continuously prosecuted in successive years. In the intervals he spent his holiday sometimes in doing a little home-geology in this country, sometimes in making a bold dash once more into the rocks of the Continent.

The home journeys were almost wholly geological, having usually for their object the investigation of some point connected with *Siluria,* including perhaps a week with the British Association and visits to old friends in the country.

A few of them were undertaken for the sake of seeing the field-work of the Geological Survey, and making the personal acquaintance of the officers under his command. These took place shortly after his appointment as Director. They were not repeated in later years.

Thus, early in July 1856, Murchison betook himself into Gloucestershire to see some of his old Silurian haunts. Mr. Ramsay joined him, and some time was spent by them among the Silurian and Oolitic rocks of the Tortworth district, where they enjoyed the hospitality of Lord Ducie, who accompanied them in their excursions. The journal of this time, wholly geological, seems to have been kept chiefly with a view to the new edition of *Siluria*. Among the Cotteswold hills he says, " We made various excursions in the range of the Lower Oolites, and were accompanied by a very intelligent person who had been in business in Cheltenham, and had quitted it for the hammer. This was Robert Etheridge. Judging from his celerity, his quickness in finding shells and naming them, and in drawing sections, I said to Ramsay, ' This is the man we must have to put our Jermyn Street Museum in order.' "[1]

Working his way northwards over some of the old battle-grounds on which he had won his spurs,—May Hill, Llandeilo, the Towey, Dynevor Park, Stoke Edith, Woolhope, Llandovery, and the rest,—the Silurian knight, accompanied by Lord Ducie, Ramsay, and Aveline (one of the senior officers of the Survey Staff), saw once more, and with

[1] Mr. Etheridge, whose merits were already known to Lord Ducie, had been asked by his Lordship to meet the geologists at Tortworth. He was soon after appointed Assistant Naturalist to the Geological Survey; subsequently, on the resignation of Mr. Salter, he became Palæontologist, and since that time has gradually risen to hold a foremost place among the palæontologists of this country.

more critical eyes, some of his earliest sections. Making a great circuit through that interesting region, the party returned to Cheltenham in time for the British Association.[1]

For many years the Geological Survey had been at work in Ireland, a considerable tract in the south and south-west of the island having been mapped and published. In his capacity of Director-General, Murchison arranged to visit his Irish colleagues this autumn. Hence, immediately after the close of the Association-meeting, he started for Dublin, whence, after some preliminary inquiries into office and Museum work, and a dinner or two at the Castle and elsewhere, he set out for the field to make the acquaintance of his staff, and see for himself the nature of the ground, and the condition and progress of the mapping. The weather, however, proved most unpropitious. Storms of rain and driving mist shrouded the hills and tore up the surfaces of the streams and lakes into foam and spray. In the face of such obstacles the Director forced his way southwards and westwards, taking Kilkenny and Limerick on the way, into the far promontories of Kerry. Porphyries, grits, Wenlock fossils, Old Red Sandstone, unconformabilities, and sections innumerable, contest with the elements for prominence in the records of his journal. It was pleasant to get to Muckross Abbey and enjoy a couple of days of rest. At that place he writes :—" Driven back by drifting rain to this most lovely spot, where I am living in beds of rhododendrons and every sort of beautiful plants on the Lake of Killarney.

[1] Professor Ramsay adds the following note:—" He and I lodged together at Cheltenham. He found his own old nurse there—the nurse of his infancy—and gave her £10. He had not known she was alive."

The reflection of the groves of arbutus and banks of the richest ferns in the pure and still waters of the stream, which flowing from the upper to the middle lake, encompass an isle, is quite marvellous. Yet all these glorious glimpses of nature have been brought out in full effect in water-colours by Mrs. Herbert, whose sketches of these scenes, as well as of the Alps and Italy, fully entitle her to be the lady of such a paradise. Her clear and decisive colouring, and her faithful delineation of every natural feature (true rocks in the foreground as well as in the distant outline), place her very high indeed in my estimation. How I thank her for having induced me to linger on one day more in this en-chanting place ! "

But the charms of this delightful retreat could not be carried about the country, and he had now had enough of the " roughing" under which alone Irish geology can be properly worked out in the wilder regions. So, bidding adieu to Killarney, and cutting short the rest of his programme of inspection of the Survey work, he set out for Dublin " and the horrors of Morrison's Hotel." After less than three weeks in Ireland he was glad to find himself once more in Wales.

The general results and impressions of this first trip were at once communicated in a letter to Professor Ramsay, as follows :—

" MY DEAR RAMSAY,—I had both your letters in Ireland, of which dear land I took leave this morning, believing it was no longer necessary for me to go poking into the holes and corners of Galway, where I have already seen the Silurians, so I sent Jukes and Kelly thither on a reconnaissance.

" . . . With the exception of these oases, far far aside, I really must declare that the geology of Ireland is the dullest

('tell it not in Gath') which I am acquainted with in Europe.
If St. Patrick excluded venomous animals he ought to have
worked a miracle in giving to the holy isle some one good
thing under ground.   But no! everything has had a curse
passed upon it.   There are as good Cambrian rocks as need
be, but they are all like the Longmynd, and won't give good
slates.   Then there are as good Carboniferous Limestone
and Millstone-grits as any in Scotia, but it is pitiable to see
the miserable small packets of broken culm at intervals of
scores of miles, which are dignified by the name of Coal-
measures.   Then as to mines it is *nil*, except what used to
be called the curse of the miner (pyrites).

" Jukes is a fine energetic fellow, and I made the
acquaintance of all his men (inspecting their work), who are
really good hard-working youths, who can stand a life no
Englishman would tolerate.[1] . . .

" I am now convinced that we must have more workmen employed in the English Survey, and specially in our
coal-districts, or some of these days we shall be blown up
by the Parliamentaries." . . .

No sooner had he got back to Montgomery and Shropshire than he set to work at once upon his Silurian rocks,
taking with him Richard Gibbs, the fossil-collector of the
Survey, whose sharp eye and stout hammer could turn
fossils out of a rock in which nobody else perhaps would
have found anything.   Hence a postscript to the last quoted
letter runs thus :—

" *Friday evening.*—Just got back from the Stiper Stones,
where I had a good tramp with Gibbs.   It is well I went to

---

[1] On this letter Mr. Ramsay remarks :—" When he came back he
said to me, ' Catch me going to Ireland again ! ' and he kept his word."

see the things *in situ*, for by persevering I got fossils all the way down to the Stiper Stones, and under them too. It is a perfect fossiliferous descending series, with *Graptolites, Trilobites, Orthoceratites,* and *Orthidæ,* as well as *Lingulæ,* both great and small, and is so irrevocably dovetailed into the series that no man alive can separate them in the field whatever Salter may do in his closet. I shall now adhere, with infinitely greater pertinacity than ever, to the original Silurian base, and standing on the Stiper Stones will defy all the world."

In later years the Director-General now and then spent

The Western Face of the Stiper Stones.

a few weeks among the slates of Skiddaw, or tracing the development and boundaries of the red sandstones of Cumberland, Westmoreland, and Lancashire. Professor Harkness there joined him as his companion, and they conjointly added some new and interesting particulars to our knowledge of the palæozoic rocks of the north-west of England. When on one of these excursions Murchison wrote some gossiping letters to his friends Sir William Denison and Sir Henry Barkly, from which some extracts will take us best into the current of his work and thoughts.

To the former he writes (16th August 1863) :—" On a Sunday afternoon, when far away from the smoke and noise of the metropolis, in which I have been presiding, chattering, eating, and drinking for the last eight months (barring a little pheasant-shooting up to February last), here I am in the middle of my Permian rocks in Lancashire. . . .

" I tried hard before I left town to get some honours of the Crown from my good friend Lord Palmerston for the men of the Nile, Speke and Grant, and though I failed for the moment I am sure the right thing must be done. (See the postscript to my Address.)

" I never expected to see my country drifting again into a war for an idea. We did so in the Crimean War, and as Mr. Bull required to be *let blood* after so many years of stagnation, I suppose that, folly as it was, the thing was inevitable. But as the only result of that war was to raise France egregiously, and above all in her maritime condition, and almost to elevate her beyond us, I could not have conceived that we should have been on the point of still further raising her and advancing her to the Rhine on account of the Poles—a people who have never known, and will never know, how to govern themselves. Mr. Bull has strangely changed from his old character if he thus Quixotizes.

" We are all dead sick of the brutal American struggle. I have always wished for the South, because they fought like noble fellows for their independence."

To Sir Henry Barkly :—

" Our meeting at Newcastle [British Association] was a very good one. I was, of course, well satisfied, inasmuch as my section of Geography and Ethnology was the most popular by far of the divisions of our Parliament of Science.

" I have effected a considerable change in our geological maps of England in this recess. There is always something to be done, even at home ! If you look at any one of the geological maps of England, including my own little one, you will see that in Westmoreland and Cumberland, all the valley of the Eden, up to Carlisle and round into Lancashire, by the coast of Whitehaven and Furness, is laid down as New Red or Trias. Now I have demonstrated, in conjunction with Mr. Binney and Professor Harkness, that all this region is Permian. I have further shown, what these two gentlemen were at first indisposed to admit, that on the western side of the Pennine chain the Permian group exhibits a large mass of sandstone, superior to the Magnesian Limestone (near Bees Head), which is also an integral part of the group, as in tracts of Germany (see *Siluria*).

" Harkness and myself also determined a fact of some importance to the amplification of my Permian group, viz., that the hematite iron ore of Cumberland, which lies in cavities of the Carboniferous Limestone, is a part of the Rothliegende or Lower Permian. This fact is quite new.

" When I look round the world I cannot help saying how grateful we ought to be who live at home and ease in these islands. With such horrors as are going on in America, where the mob rules supreme, or where their President is as odious a tyrant as imperial Rome could have produced, it is wonderful that such men as Everett and Agassiz should write to me from Boston as if nothing were occurring that would not soon pass away, and as if the great republic would soon be one and indivisible ! "

The reference in the foregoing extract to the Newcastle meeting of the British Association suggests here an allusion

to the fact that at the meetings of that body Murchison
still continued to take a prominent and useful part.   We
can hardly count them properly with his holiday, but coming
as they did in the time during which he escaped from London life, they may be briefly noticed in this chapter.   His
geographical ardour, which, as before remarked, led to the
creation of a special geographical section at the Association
meetings, eventually made that section one of the most
popular of all.   He had a paternal interest in it, and used
in a half-jocular, half-serious style to boast of its attractiveness, and of the way in which it had eclipsed the other
sections, even his own old favourite " C."   With a good
" lion " in the shape of a Livingstone, a Speke, or a Baker,
he was sure to fill his meeting-room to overflowing, and
knowing this, he did his best, when in the chair, to secure
the attendance of some such explorer, or, failing him, of as
many geographical notables as he could induce to come.   His
enthusiasm in behalf of the geographical element of the
British Association was never more ardent than on the occasion of the Newcastle meeting in 1863, as the following
extracts from his notes to Lady Murchison will show :—

" DEAREST C.,—" We had a right capital day yesterday.   Grant filled the section to repletion.   There were
1200 persons in the Assembly Rooms, and he performed
twice as well as he did in London.   I am anxious that you
should read the accounts given in the *Newcastle Express,*
which goes to-day, because every word I said is well reported, and I always wished my anecdote about Grant's
gallantry in India to be well put forth. . . .

" To-day we had a no less successful day—beginning
with a paper by Lord Lovaine on newly discovered pile lake-

habitations in Wigtownshire, which I got Lyell to attend, and at which he spoke well.

" Half of this day was devoted to the Geological, where I held forth at some length on my Permian rocks of the west of England with Harkness, and on the reptiliferous sandstones of Morayshire. Yesterday I proposed we should meet at Bath, and Lyell for President, and it was carried by a large majority.

" Yesterday, also, they gave us a dinner, at which I had to propose their healths, with all due estimates and comparisons of past and present Newcastle.

" I go to Alnwick Castle on Thursday, and write to me thither. The Duke kindly wrote to me to bring Speke and Grant. . . . I have been asked to Corby Castle, and numerous places, including Clumber (Duke of Newcastle), so that if my forces hold out, and I make my tour to the Highlands, God knows when poor Pincher,[1] to say nothing of my loving wife, will see me.—Ever yours,   ROD."

" I told you that the Durham boys had asked the ' Old Boy' to get them a holiday, and I wrote to the head-master for it. . . . . Dr. Holden came up to me after the meeting, and congratulated me, and told me he had given the boys a holiday for the day, at my request. The Section thanked me warmly when I took leave. We beat all the others in popularity."

Before passing from the British Association we may take some further extracts, showing the meeting under a very different aspect. In the year 1864 it met at Bath, under the presidency of Sir Charles Lyell, and with Mur-

---

[1] A favourite terrier of Lady Murchison's later years.

chison as the leader of the geographers. The Geographical Section mustered strong, especially in African travellers, Livingstone, Burton, and Speke being there. But before its close the meeting was thrown into mourning by the sudden and distressing death of one of these great pioneers. Murchison writes of the meeting and the tragedy thus :—" Livingstone was the greatest lion of the meeting, and was with his charming girl Agnes staying in the same house with myself. Speke came in from a friend's house in the neighbourhood to reply to a paper by Captain Burton, which he knew would be antagonistic. That paper being deferred, he left our meeting before two o'clock and rode back again to the country. There, taking his gun to kill a few partridges, and accompanied by a young friend, he met with his death by incautiously pulling his gun at full cock after him in getting over a stone wall.

" Livingstone and I went from Clifton into the heart of East Somerset to attend the funeral. It was indeed a very touching scene. There were met together the great South African explorer Livingstone, and Grant, the companion of Speke, with myself, the historian and profound admirer and friend of these noble fellows. The funeral procession, with all of us on foot, proceeded from the pretty little parsonage of his brother, the Rev. B. Speke. The country people lined the roadside and hedges as we moved down to the little church, and then Grant placed an *immortelle* on the coffin of his leader."

The event is again touched upon in a letter to Sir William Denison :—" Since I received your letter of the 29th July, I have been going through a good deal of purgatorial business for the British Association. My Section E, or

Geography and Ethnology, was eminently successful, although our proceedings were necessarily clouded by the catastrophe of poor Speke's death.

"I have suggested the erection of a monument to his memory, and hope we may get enough to have an obelisk or something appropriate raised. I attended his funeral, and it was a very touching scene to see all the country folks out and the poor father weeping as he followed the coffin of his daring and intrepid son to the grave. Grant, his companion, came from the Highlands expressly for the purpose. These explorers have done so much honour to the Indian army that I hope their companions in arms will respond to my call (see my letter, *Times*, 27th September)."

The home holidays were sometimes given to amusements not by any means scientific. For example, in the autumn of 1858, he paid a visit to Lord Derby, and there saw English statesmen in a light quite new to him. From the full notes which he made of this visit, he evidently considered its details worthy of preservation. The Social Science Congress, with Lord Brougham at its head, had met at Liverpool, and after its sittings, some of its members were invited by the Tory Premier to Knowsley, where also Murchison arrived about the same time, and found, that among the guests were "Lord and Lady John Russell, with their daughters, and Lord Carlisle, both of them Whig statesmen, who have been spouting the whole week on social science with Brougham and his allies at Liverpool, forming now part of a happy family with Derby and Walpole of the present Government."

After dinner the party, sixteen in number, set to at high jinks in the drawing-room. "The games began by a name-

less one, in which we all stood round in a circle holding our
hands on a white band—Lord Derby beginning as the unfortunate man in the middle, whose play it is to hit the first
person's hand he can, to avoid which the hands of the company are kept in perpetual motion on the band. For the
first few rounds I observed that the statesmen were oftenest
in the ring; but the old geologist, and every one, had his
turn, though I think Lord John was most frequently *in.*
This was really as good exercise as a Highland reel."

The next game consisted in sitting back to back, getting
up as music began, walking round and round the chairs, and
dropping into the nearest one when the piano ceased, there
being always one chair removed at each round. Then came
the time-honoured " Post," when, as each one chose his town,
the geologist, fresh from the wilds of Assynt, took Inchnadamff, the name of his inn there.   " The master of the
jinks," he says, " seemed particularly anxious to nail old
Inchnadamff, for the post went oftenest there, though in
reality it only goes twice a week ! At first I got off by
quiet boggling, but on another occasion, when rushing, I
was caught.   Much fun."

" What a lesson do these frolics read to those who think
that political strife is not compatible with real *bonhommie*
and private friendship ! How would foreigners above all
stare at seeing these gambols of a ' happy family' ! How
impossible to realize in France a game in which Guizot
and Walewski, Lamartine and Persigny, should all be
amusing themselves together like good fellows !"

On one occasion Murchison's autumn ramble, which had
originally been designed for Highland geology, was pre-
vented by bad weather. Having got as far as Edinburgh,

and perhaps being somewhat depressed in spirits, he gave
up his plan for doing more geology, and took instead to the
very different and rather mournful occupation of revisiting
the scenes of his boyhood and youth. At Edinburgh the
recollections of sixty years back came fresh upon him—the
recruiting party, the balls and dinners, the bright young
faces which once captivated the heart of the fledgling en-
sign. Older still were the reminiscences of his mother's
care. To the last he cherished her memory as one of the
most precious possessions of life. With pious steps he now
seeks out the places to which she took him when a boy, but
finds them so changed as to be hardly recognisable—Fisher-
row, Lasswade, Rosslyn, Peebles, all grown and modernized.
He makes a pilgrimage to Dryburgh, to the tomb of his
friend Lockhart, who, though two years his junior, had pre-
deceased him. At Raby Castle, once the headquarters of
fox-hunting, he finds the old Duke, his former sporting
chief, in enfeebled age. At Rokeby his quondam host had
given place to a younger squire, and other changes had followed. Barnard Castle, his own home in the early years of
his married life, now boasted of a railway station. On a
site which he remembered to have been marked by only a
single farm-house, he sees the large and populous " iron
town " of Middlesborough ; at Hartlepool, where he used
to shoot among marshes, he comes upon a range of docks
crowded with shipping ; at Stockton he finds a forest of
smoky chimneys replacing the quiet little old town where
he " danced in 1810 with Miss Milbanke, shortly afterwards Lady Byron." In the same neighbourhood he visits
Mrs. F—, a lady of 74, who more than half a century
before had been his sweetheart !

The foreign tours undertaken by Murchison during the last fifteen years of his life had mainly for their object the recruiting of his health by rest and change. But in most cases the very sight of the old rocks among which he had once worked so hard was enough to re-invigorate him, so that, though he came to them somewhat of an invalid, he quitted them with renewed vigour of body and buoyancy of mind.

A tempting prospect opened in the summer of 1857. Having seen the succession of palæozoic rocks in Europe, the author of *The Silurian System* was naturally desirous to examine personally the remarkable development of the same rocks in North America, where they had been so successfully worked out by Hall, Logan, and others. There was to be a great congress of scientific men at Montreal, and the learned Societies of this country were invited to send delegates to that meeting. Murchison wished to go as the representative of the Geological Society. He had even decided his plans, when they were peremptorily broken up by his medical adviser.[1]

Nevertheless an autumn tour of some kind was essential, and though he had to content himself with a less ambitious programme, he determined to go abroad and set before himself a task which would give point and interest to his travel. In the end he arranged a ramble through some of

---

[1] The Geological Society of London having resolved to send a delegate to the gathering at Montreal, Professor Ramsay was first proposed on the supposition that Murchison would not go. When it was found that the latter did wish to undertake the journey the arrangement was changed. Mr. Ramsay writes:—" Murchison consulted Sir Henry Holland in the hall of the Athenæum when we were together. Sir Henry looked into his eyes, and at the white ring that encircled the iris, and said he must not go to America." The first choice was therefore adhered to, and Mr. Ramsay was selected, and went accordingly.

the best tracts of Germany for the exhibition of the Permian rocks—a geological system to which, despite his paternity of its name, he had paid no special attention since the great Russian tour.   The plan adopted was thus described :—

" 30*th July* 1857.—MY DEAR M. BARRANDE,—After having arranged everything for a voyage to Canada and the United States, commencing with the great meeting of *savans* at Montreal, I have changed my plans, seeing that my doctor rather prescribed for me rest, and the quiet amusements of Germany.   The truth is, I have been working too hard this year, whether as President of the Geographers, Director of the Geological Survey, or Trustee of the British Museum, etc., and now having printed my anniversary Address to the Geographers (114 pages), I embark on Sunday night, 2d August, for Antwerp.

" Thence I shall ramble on towards the Thüringerwald, and establish myself there for ten days at the baths of Liebenstein, of which I am very fond ; because one can get there pure and cold water, perfect shade, and capital Permian rocks.   So write me a note to say if I shall find you at Prague in case I should bend once more (and always with great profit) towards the Klein-Seite.[1]

" The brave Peach has again discovered fossils in the crystalline limestone and quartz rocks of the north-west Highlands of Scotland—species identical with those of the Calciferous Sand-rock of North America.   Is this not beautiful ?   I am enraptured with it.   Poor Hugh Miller conceived hypothetically that these rocks represented the Old Red Sandstone, and Nicol has recently suggested that they

---

[1] M. Barrande's address at Prague.

are nothing but the coal formation changed into quartz rock and mica-schist!—Ever yours,   Rod. I. Murchison."

His companion in this tour was Mr. T. Rupert Jones, Assistant Secretary of the Geological Society—a name now familiar wherever palæontology has made its way. Keeping in view the examination of the sections and fossils of the Permian rocks as their main business, the travellers passed over a large area of the Continent. For after ascending the Rhine, and lingering at Liebenstein and Prague, they went on to Vienna, and then, wheeling round by Breslau, struck north for Berlin. Thereafter a pleasant time was passed in the Harz, whence a leisurely progress by Cassel and Frankfurt took them to Bonn in time for the annual gathering of the German Naturforscher. Lastly, making a run into the Lower Rhine province, they turned their faces homeward, and got back to England after an absence of about two months.

The journal of the tour is as usual almost wholly geological, and its scientific details have been already published. From its pages, and from the letters of the time, some extracts of a more generally interesting nature may here be culled.

At Liebenstein a courtly and flattering letter from Humboldt was found by Murchison awaiting his arrival, in which, among other phrases, the writer stated his opinion that "celui qui de tous les géologues vivans a embrassé la plus vaste sphère des connaissances précises sur la structure de notre planète, c'est Sir Roderick Murchison. C'est l'opinion que je proclame." And the courtier conveyed further a hope that the geologist would visit Potsdam.

This in due course Murchison did, paying his respects once more to the King, and spending hours in multifarious gossip with Humboldt. Among other remarks of the learned Prussian, the journal records that he " did not care sixpence for the ' Reseau Pentagonale ' of É. de Beaumont.[1]   In alluding to his present onerous position, in standing between the King, with his decorations and pensions, and a large host of correspondents, he assured me that the mere postage of his 2500 letters per annum now came to £150, which, as he said, would educate two youths. Some of the letters he had to answer were most absurd. Thus he brought out one from a M—, which in about thirty pages endeavoured to show that all the Bible could be explained by certain figures —6600—and was full of Greek and other quotations ; also letters from pious young ladies, who sought to have the privilege of closing his eyes when he died. He spoke of his manifestly decaying strength. Though he could stand for hours, his limbs began to fail in walking, ' But there is nothing to be complained of,' said he, ' for if I live till the 14th of this month I shall be eighty-eight.' "

The next extracts take us into the midst of the gathering of the Naturforscher :—

" *Bonn, September* 20*th*.—Reached this place after a very fine day on the Rhine, during which our steamboat passed through the fleet of *savans*, who, in their large steamers,

---

[1] This was the fanciful network of lines drawn over the face of the globe by the ingenious Frenchman, to mark what he considered to be the direction of the different contemporaneous systems of elevation. In later years, not long before his death, Humboldt, in writing to Murchison, spoke of De Beaumont thus :—" Élie de Beaumont fait le Kepler sans découvrir des lois. Il perd son temps à construire le grand Pentagone et croit toucher le clavecin de la Nature ! Ce que c'est que d'exagérer une idée sur laquelle on est à cheval depuis 25 ans. C'est de l'ennui en 3 volumes digne du Sénat Impérial ! "

with a hundred flags flying and pateraras firing, were at
that moment saluted by the population of Coblenz, who
paraded on the quays.   Old Nöggerath's speech and the
Princess of Prussia's reply.—On arrival here found Dechen
just returned from an expedition to the Siebengebirge, and
had tea and supper with him and his friends.   The return
of the steamer at ten o'clock at night, in a bright starlight,
the firing of the guns, and the reflection in the water of the
fireworks of the town, were very fine. . . .

" 21*st.*—Adjourned to the old theatre, and when Nöggerath had arranged all Mitglieders on one side and Theil-
nehmers on the other, the business of the day began—that
of selecting the next place of meeting.   Emms on the Weser,
Carlsruhe, and Düsseldorf competed.   Nöggerath made a
series of jokes and puns with his arch gravity and stento-
rian lungs ; one of which was about the division of Germany being no longer that of North and South, but, as
Leopold von Buch had termed it, the land where they drink
wine out of small glasses, and where they drink it out of
large ones.   Another was that in speaking of an individual
he called him Schmidt, and being corrected by being told
that Schwarz was the man's name, he roared out, ' Das ist
alles ein : der Schmidt muss immer *schwarz* sein ! '

" After pleadings and speeches, the best of which was from
Von Carnall, who, supporting the claims of Carlsruhe, denied
the soundness of the division of Germany by the larger or
smaller glasses of wine, and said they were now united, and
one Vaterland, in drinking their good old beverage of *beer.*
This hit was immensely applauded, and warmed the cockles
of their German hearts, and specially of Naturforscher and
medical men who drank nothing else.   The palaver ended

1857.] *PROFESSOR NOGGERATH OF BONN.* 277

by Carlsruhe being unanimously selected, and by Eisenlohr, the mathematician and astronomer (who was one of my Southampton flank in 1846 at the British Association), being chosen President—a jolly good-humoured fellow, who can hold a gallon of beer and will make a popular President. Then came the dinner at one o'clock. I dined at our hotel, the Star, where about 150 to 200 sat down—Dechen, De Verneuil, Abich, Koksharoff, Hermann von Meyer, and a fair sprinkling of good geologists. This lasted two hours and a half.

" Then followed an excursion to Rolandseck, to hear Nöggerath give a discourse on the chief volcano of Roder Berg, about one and a half mile inland from the Rhine. The old cock led the motley group of ladies and gentlemen, and very few geologists,—in all a mass of about 300 to 400,—through the woods, slopes, open fields, etc., and then stepping across, he said, ' We must mark the potatoes before we get at the cutlets;' and he took us across the crater, about three quarters of a mile in diameter, and explained to us how the wind must have blown from north to south during the eruption, of which there were evidences in the fact that *lapilli* and scoriæ twice alternated with loess and Rhine pebbles.

" We descended after sunset, and walked a mile and more to the other station, where, after waiting three quarters of an hour, De Verneuil, Daubrée, Jones, and myself, jumping into a first-class carriage *coupé*, escaped, and thus preserved ourselves for the rewards of the evening, all dead tired. Old Nöggerath, however, never knocks under, though he is in his seventieth year.[1] . . .

[1] Professor Ramsay remarks regarding this veteran :—" He still lives,

"Another great dinner at the 'Stern,' with champagne and speeches. To Nöggerath's toast of 'The Foreigners who had honoured the Meeting,' I replied in French for the English, followed by Élie de Beaumont. Evening at Dechen's again.

"Thursday, our last day of meeting. Morning at Poppels-dorf; long communication of Abich on the Caucasus, and on his view of dislocation; Dumont's map of Europe exhibited. We made an excursion under the leadership of Dechen to inspect the basalt of the Siebengebirge."

Shortly after the return to England a report of progress was sent to Sedgwick :—

" October 15th.

" MY DEAR SEDGWICK,—You may have learned that I gave up a trip to North America, on which I had set my heart, because my state of health and nerves would never have stood the excitement and wear and tear of Jonathan's hospitalities.

" I went, therefore, to Germany, and to many of our old haunts and some new ones, accompanied by Jones of the Geological Society, whom I took as my aide-de-camp, and a capital staff-officer he proved. I have come here well and strong, and hope to hear you are the same. My wife is living, and will live, at St. Anne's Hill, the residence of Charles James Fox, till the 1st March, as London does not agree with her, and I go and come thither and hither.

" My great object in Germany was to see every good natural section of Permian rocks, and to commune with the

and is a fine type of the old school of German Professors. In 1860 he told me that after the Princess-Royal's marriage, when Prince Albert brought the Queen to that part of Germany, the Bonn Professors went to be presented to Her Majesty, and Prince Albert took his old master (Nöggerath) by the hand and said, ' Come and be introduced to my *wife.*'"

best men who had written thereon, and I have succeeded very much to my own satisfaction. . . . I have thus got all the German Permian in my pocket. Feeling that much is to be done in the English Permian, I intend to go down for a week to look again at several sections, which, in the north of England, have been laid open by railroad cuttings. . . . You will much oblige me if you will send me your opinion, and a hint or two as to the spots where I am most likely to see clear data. . . .

"Our old friend Nöggerath played his part stoutly and heartily, and as our dear old friend Dechen received every evening at supper, and De Beaumont and De Verneuil came from Paris, with Deville, Hébert, Daubrée, and the French geologists, I enjoyed the meeting much. Dechen and your old allies begged me to convey to you their kindest remembrances.

"De Beaumont spoke to me about the vacancy in the Institute caused by the death of Buckland, and then I told him of the melancholy addition to the list of poor Conybeare. 'One of the vacancies,' he said, 'we must of course fill up in our list of correspondents, and it is my intention to propose Professor Sedgwick, who will, I hope, be unanimously elected.' I told him that I had already conveyed my sentiments on the question being put to me by a mutual friend, and that there could be no hesitation in bringing you in before any of our English contemporaries, and that according to your merits you ought to have preceded me.

"I hear that D'Archiac is to be the new Professor of Geology in the Jardin des Plantes, *vice* D'Orbigny. Send me good accounts of yourself.—Yours very sincerely,

"ROD. I. MURCHISON."

About three weeks later, having meanwhile carried out his intention of visiting the north of England, he again reports to the same friend :—" I returned from my Permian skirmish on Monday night, having explored many of your old beats, and, I am happy to say, with a sincere admiration of your old and most excellent memoir. Your letter was a full proof to me that your memory was anything but an ' old rotten fishing-net,' for I never received from you or from any one a more clear synopsis of all that constituted truly the British Permian. . . . I zigzagged across most of your old sections, and admired them all."

Occasionally the Director-General did not wait till autumn for his holiday, but would escape about Easter for a week or two into France, not, of course, to do any field-work, but to have a geological gossip with De Verneuil, Élie de Beaumont, and other old friends. In the spring of 1860 he paid a short visit of this kind to Paris. Sauntering through the boulevards, and contrasting them and the open gardens with the narrow ill-paved streets of 1814, he could not refrain from admitting that the Emperor Napoleon, whom he never could forgive for provoking the war with Russia, had at least done good to Paris. At the same time he drew a contrast between what was possible in France and in England. "We free and insolent islanders," he remarked, "can never by any possibility have really fine public buildings, or well-regulated and attractive public places, because the hand and mind of an autocrat of taste are wanting. Yet we have one public garden as fine as, if not finer than, anything in the world, and that is Kew. And why ? simply because my friend Hooker is really autocrat there, and, doing every-thing he wishes, is no more stinted in money grants than the Emperor Napoleon. Kew Garden on the one hand, and the British Museum on the other, are the two great establishments *only* in which we beat all other nations, but the last mentioned, and much the most ancient, is said to be doomed to dismemberment. 'Credat,' the shade of —— 'non ego !' The Parliament will step in and stop the Vandalism. This, at all events, is one of the uses of an unfettered House of Commons, for ignorant as they are in matters of fine art and taste, they know what the British public appreciates, and will vote accordingly. The British Museum grew up like our constitution, and, like it, no foreign kingdom has anything so grand."

" *Easter Sunday.*—The sun continues to shine gloriously and the Madeleine and all the churches are crowded to suffocation. The clergy have certainly exercised a great influence over the present generation, and I see a prodigious change in the French organ of veneration." . . .

" *Easter Monday.*—I prepared a written article on my Highland campaigns, and read it at the Institute after De Verneuil had corrected my phrases. Élie de Beaumont and Florens, *Secretaries*, and Milne Edwards in the Presidential chair. Cordier (84), fresh as a man of fifty ; Biot, of the same age, very decrepit ; Valenciennes, much aged and bent ; Ad. Brongniart, grey and oldish ; D'Archiac, spruce, strong, and energetic, but becoming grisly ; Chevreuil (75), out of his bed and gay as a lark ; De Senarment, a fine open-spoken black-haired fellow, etc.

" Dined with De Verneuil at the Club Agricole. Long conversation on England and our policy. The general feeling in France is against the commercial treaty, to which time and development will alone reconcile them. Hence its introduction at this critical juncture, when many other causes of irritation exist, serves to fan a warlike flame. The country is essentially money-making and prosperous ; but so dislikes the war now, that when the Emperor started for Italy every one was against his going to fight for such *canaille*. 'But,' said my friend, 'if war should burst out under existing circumstances between England and France (though I should deplore it, and believe it would go far to ruin both nations), rely on it the mass of my countrymen would rejoice in it, for they believe that the hand of England is everywhere meddling to the disadvantage of France.' "

The two friends and fellow-travellers, quitting politics with the gaieties and discussions of Paris, made their way to the district of Amiens and Abbeville, which had recently come so prominently before the world in connexion with the question of the antiquity of the human race. They looked at the gravel, saw the way the flint tools lay in it, and visited M. Boucher de Perthes, who, in his old age, suddenly found himself and his museum famous. De Verneuil then returned to Paris, while his companion got back to London.

The summer of 1862 proved to be a trying one to many of those who had duties in connexion with the International Exhibition. Murchison in the end precipitately fled from London, and took refuge at some Bohemian baths, from which he sent the following account of himself to Sir William Denison :—

" *Marienbad, Bohemia, August* 28, 1862.—MY DEAR SIR WILLIAM,—As I may never have more time at my disposal than at present, when I am just finishing off my cure with

the Marienbad waters in Bohemia, and have sufficiently
stimulated my sluggish liver, etc. etc., I beg to thank you
for your last letter, to which I fear I never replied.

"The last season in London was most antagonistic to
all correspondence on my part, inasmuch as I had the
misfortune to be the Chairman of Class No. I. of the
International Exhibition, and had, with my jurors, to arbitrate
among nearly 3000 exhibitors. Moreover, they would
insist in my re-occupying the presidential chair of the
Geographical (*vice* Lord Ashburton), and what with other
occupations of various sorts, and loads of dinners, public
and private, I was somewhat used up.

"Here I have been drinking and toddling about among
Austrians, Prussians, Poles, Hungarians, and a few of our
own countrymen and women for the last five weeks. Now
being furnished with the newest geological maps of Northern
Bohemia from the Geological Institute of Vienna, I am
about to take my hammer in hand, and revisit, for the
third time, the magnificent Silurian basin around Prague,
which has been admirably laid open by a new railroad from
Pilsen to that city.

"I have also a strong desire to ascertain if, in the
crystalline centre of Germany, there are not gneissose rocks
of the same remote antiquity as those which I have developed in the Western Highlands of Scotland, and which
there underlie all the fossiliferous strata. (This is quite new
in British Geology.) I also wish to make out some additional features in my Permian rocks, and these things accomplished, I shall traverse to Brittany, and see if in the environs
of Cherbourg I cannot also discover some of my 'fundamental gneiss' (for there are gneiss rocks of different ages).

" Here, embosomed in pine forests, we have splendid large crystalline granite, through which the gaseous and saline waters, with iron (the remnant of a former intense volcanic energy), bubble up. Certainly there is no watering-place better calculated to suit a variety of persons amid these several wells, each of which differs from the neighbouring sources by containing more or less of salt, iron, and carbonic acid gas. We are all obliged to use the carriage of the apostles, and there are many miles of shrubby walks all odoriferous of pines. Whilst I write, I go from time to time to my window to mark the sad progress of a fire in a village some six miles off, as seen in the undulating lower country. This is the second village burnt down since I came here.

" We have a pleasant English circle, which Lord Clyde was to have joined last week, but his companion, my old friend Count Strzelecki, came without him, and is in the same house with me. . . . When I came, and for my first fortnight, the best of the Austrian Generals, Feld-Zeug-Meister von Benedek was here; he who smashed the Piedmontese army at Solferino, and was called back from his pursuit to break his heart by the hurried peace which his young Emperor made with the crafty Louis Napoleon. Benedek is quite confident that the French would have been brilliantly beaten if the war had gone on. He utterly denies that the fortresses in the Quadrilateral were unprovisioned and unprepared, and he smiles at the idea of long shots of artillery deciding great general actions. I once heard the great Wellington say the same thing. Benedek commands the Austrian army in Italy, nearly all of whom, with himself, are Hungarians. There is no doubt that it is

a noble army, and I heard a French officer say last spring,
at Count Flahaut's table in London, that it was the finest
army in Europe, though it had been ill-commanded. If
they have another burst when commanded by Benedek, the
results will be very different.[1]

" By the death of an old aunt, I have become the holder
of rupee bonds, and, as they give me a good interest, I shall
hold on, as such as Elgin and yourself are our guarantors in
India. Pray make some new geological or geographical
observations. I cannot solve your meteorological guesses,
but I am always much interested in your descriptions of
what you see or know. I shall be at my post in six weeks
for the opening of the Royal School of Mines, for that is the
new title of the Jermyn Street *locale,* including the Geological Survey, Museum of Practical Geology, Mining Record
Office, etc. etc.—Yours sincerely,	ROD. I. MURCHISON."

Of the subsequent ramble into Bavaria, and thence westward through France, little need be said here. The renovated
geologist felt that he must be busy among the rocks again.
But though this desire returned with much of its former
fervour, it was not seconded, as it used to be, by the same
indomitable activity of body. Besides, he had not come
prepared for any really serious work, so that his geologizing
was, to use a phrase of his own, more " to keep his hand
in" than with the view of eliciting anything new and
important. At Munich, among the picture-galleries, the
connoisseur and critic in fine art re-appears, recording his
judgments in his note-book very much in the same style as

[1] They certainly were different, the difference being sadly against poor
Benedek in the Austro-Prussian campaign of 1866.

in the early days at Rome.   Music, too, receives its share of
attention in the same pages.   Thus :—" At nine o'clock to
the great church to hear morning service and a mass of
Mozart; most touching and sublime ; produced more good
within me than any amount of preaching and damnation ;
great devotion of an immense crowd.   The *Miserere* and
Gloria both exquisite, and more touching than Handel.
Haydn's *Offertorium* beautiful.   The musicians are those of
the Court, and the best in Munich."   And when everything
else fails, the inveterate " taker of notes," in his railway
journey through the broom-scented and well-wooded valleys
of the Vosges, consoles himself with a malediction on his
travelling companions :—" Stupidity of the carriage-full of
English folks, who looked at nothing, read trifling books all
the way, and asked no questions.   One old gentleman said
it was extraordinary how the clocks differed in these countries ; ' but,' he added, ' I suppose I shall find my watch
right when I get home.'"

Rapidly traversing France, he halted at Pont St. Maxence,
to find there " De Verneuil kinder than ever, and always the
same thorough friend."   During the sojourn at Paris art
again takes a large share of the memoranda : " The Louvre
can never be visited too often.   It is now an infinitely more
wonderful assemblage of works of art than when I first saw
it in 1814, when all the choicest pictures of Italy were in it,
and had not been restored to their owners.   It is true that
some of the grandest things have gone, but their places have
been trebly filled, and the quantities of fine pictures of the
best masters is such that, in admiring, an Englishman cannot help being vexed at the poor condition in which we stand
at home.   I never admired Spagnoletto much, but there is

an ' Adoration of the Shepherds' in the large circular room
which is quite Corregiesque. I would as soon have it as
any picture in the Louvre. The completeness of the vast
assemblage is such! The whole history of art from the
earliest efforts of the Egyptians to the renaissance of the
past centuries, and thence to the present day, is all before
you. I see that the French folks set some value on Ros-
alba's pictures in their own national collection, and hence
I shall now esteem my own more."

To interrupt for a little our narrative of holiday tours
abroad, it may be mentioned that an unfortunate relapse,
caught somehow in the homeward journey from this excursion, went far to undo the refitting which the Marienbad
waters were believed to have effected; and thus, when the
geologist landed in England again, he passed at once into
the hands of Dr. Bence Jones. Late in the autumn he had
usually never found any medical treatment better than a
gun in a good pheasant cover; and having several standing
invitations, he proceeded to avail himself of them.

Proceeding first to Lord Palmerston's, he writes,—" At
Broadlands I enjoyed some shooting with the dear Viscount
and fine old Admiral Bowles. As we went into the wood a
cock-pheasant crossed me, and down fell two! Palmerston
was delighted with the shot of the old geologist. A second
bird happened to be crossing which I did not see. . . . We
had some pleasant visitors. Among others, Henry Bulwer
arrived with Turks and long pipes, etc.; was amusing by
flights, important, mysterious, and—disappeared."

Lady Murchison, not in very good health, had remained
at Brighton while her husband made his round of visits. He
was suddenly summoned to the coast again by a very serious

and what at one time seemed likely to prove fatal illness.
Among his papers there occurs a loose sheet, with some
reflections written during the suspense, and showing the
tender and grateful feelings with which he regarded her :—

"*Clermont, Norfolk, Nov. 5th,* 1 P.M., 1862.—Alas! the
telegram received calls me to my dear wife's sick-bed at
Brighton, and here I am fast bound for three or four hours
for the want of any railway train, and doubting whether I
shall catch the Brighton train at night! What a painful
state of suspense, and what a journey I have before me!
What a happy retrospect, and what a sad prospect!

"I look to her as having been my safeguard and guardian
angel for forty-six years. *She* first imbued me with a love
of science, and weaned me from some follies of the world.
She accompanied me in the three or four first years of my
geological career by land and by water ; she sketched for
me, collected fossils for me, and encouraged me onwards.
Even when I was working at my *Silurian System,* twenty
years and upwards after our marriage, she was often by my
side, and from those days to these, when unable, from feeble
health, to accompany me, she has been my best adviser,
and my infinite solace when I returned to my own fireside.
Her goodness, her deep sense of religion, and her practical
benevolence, devoid of all cant and profession, have often
made me reflect with sincere sorrow on my unworthiness of
her goodness,—on my vanity and love of the world and its
pleasures, as contrasted with her humility and true Christian
piety.

"I ought to be a much better man than I am after so
many years of so good and excellent an example before me.

"May the Great Disposer of all events have so prearranged all human destinies in a future state that I may be able to witness her heavenly abode (for that surely it must be) should such a miserable sinner be far removed from her!

"*Brighton, Nov.* 6.—A tedious and anxious journey by the Eastern Counties railroad from five till ten (Brandon to Shoreditch). Delayed by accident. Reached this at midnight, and thank God to find her rather easier, and most grateful to me for coming.

"*Nov.* 9.—Three intensely anxious days. My dear wife cannot shake off the bronchitis. Coughing all night, and incapable of eating. Nourished by beef-tea and arrowroot. Has had intense suffering. Mind wandering occasionally. Mr. Turner, the experienced surgeon and practitioner, was doing all in his power. Myself in a state of deep affliction, and oppressed with the calls on my duty to-morrow as President of the Geographical Society."

CHAPTER XXVII.

THE ROYAL GEOGRAPHICAL SOCIETY.

In the course of the previous chapters the rise and growth of the Geographical Society have already been incidentally dwelt upon. We have seen how, from its beginning onwards, Murchison had identified himself with that Society, and how, in his later years, it had gradually so engrossed his time and thoughts, that his old love, geology, could no longer boast an undivided empire over him. In watching his career, too, it is to be noticed that much as he liked the publicity and display of a Society so rapidly growing in popularity, the constantly pressing demands which its official and routine work made upon him proved to be more than at his now advanced age he thought he could satisfactorily fulfil. Hence after, among other services, procuring for the Society a royal charter of incorporation, and greatly augmenting its membership, influence, and wealth, he in 1859 resigned the chair to Lord Ashburton, though with the full intention of remaining at the Council-board, and continuing to give advice, and, if need be, active assistance.

Lord Ashburton's failing health, however, made Murchison's retirement more nominal than real. The ex-President had still to carry on a very large share of the Presidential

work of the Society, which, with its now world-wide correspondence and connexions, was growing every year heavier.
At the last moment, too, when the May anniversary of 1861
was approaching, Lord Ashburton, suddenly called away
from London by illness in his family, left the preparation
of the annual Address as an additional load to " the willing
horse." Hence, in 1863, after a brief nominal retirement, Murchison, at the entreaty of his colleagues, once
more took the chair, and held it up to the last year of his
life. It was during this later period that he kept the Society
most prominently before the world, and came himself to be
so widely known to the general public for his keen interest
in geographical research, and in the fate of geographical
explorers. Some more detailed notice of this part of his
career may therefore fitly find a place here.

Upwards of thirty years had passed since that committee
of the Raleigh Travellers' Club met out of which the Geographical Society took its rise. During that interval, though
there had been no adventurous voyages of discovery, after
the Drake and the Raleigh type, far more had been done to
extend our knowledge of the geography of the globe than
during any previous period of similar length. With this
progress of research the Geographical Society had been
honourably associated. Some of its members had distinguished themselves as bold and successful explorers ; but it
sought to reward and encourage every intrepid traveller,
whether belonging to its ranks or not, and eagerly embraced
the opportunity of publishing in its Journal, and in this way
making widely known, the narrative of his discoveries.
Having been so long and so intimately associated with the
work of the Society, Murchison naturally felt and expressed

a just pride in the services which it had been enabled to render to geographical research. Looking back through the lifetime of more than a human generation,—over the vast additions which had in that time been made to our acquaintance with the surface of the globe,—he could point to not a few which had been achieved by members of the Society, or by others who owed some at least of their success to the stimulus and assistance which the Society had given to them. He could further boast that the evening meetings, discussions, and publications of the Society had been the means of first making known to the general public the nature and value of such discoveries.

There were three regions in particular, the exploration of which had been watched by Murchison and his associates with keen interest,—the interior of Australia, the interior of Africa, and the lands and seas lying round the North Pole.

With regard to Australia, Murchison, when President of the Society, as far back as 1844, had earnestly urged the formation of settlements on the northern shores of that great continent. He had reiterated this advice in 1857, pointing out how great would be the advantages, commercially and politically, to have possession of the noble bays and harbours of that coast-line, and concluding his appeal with these words :—" Let us trust that if such a consummation [a settlement in North Australia] be obtained, the proposers of it may not be forgotten, and that it may be remembered that the North Australian expedition, now happily completed under direction of Her Majesty's Government, was a child of the Royal Geographical Society." The indomitable courage of Macdouall Stuart in forcing his way

across the continent, eventually brought about the establishment of the settlement so earnestly desired,—"an object
which," Murchison remarked, "has long been a dream of my
own, and which I rejoice to see thus realized in my lifetime."[1]

Besides the references in the formal anniversary Addresses, the President showed in other ways a keen and
kindly interest in the colonies and their future, as well as
in the exploration of new territory. At his suggestion the
Society gave a gold medal to the family of one of the Australian explorers who had lost his life in the attempt to
recross the continent. Alluding to this he writes to Sir
Henry Barkly:—

"I told you in my last that I thought it probable we
should grant one of our gold medals to the family of Burke,
and I am happy to announce to you that at our last meeting
of Council the award was made as I anticipated, and on my
own proposition, strengthened, as it was, by your favourable
opinion. It is our business to recompense the daring (albeit
rash) adventurer who is the first to accomplish an arduous
task which others have failed to carry out. We also give to
the good and intrepid King a gold watch, with an inscription. The Duke of Newcastle has promised to receive these
donations on the 26th." . . .

Again, "I am so deeply embarked in the Australian
discoveries, that I intend to open the Geographical session,
16th November, here (being once more President), by an
Australian night, when all the documents you have sent
to me will come out at least in abstract, together with
some information from Queensland. . . .

"It is really refreshing to read your excellent address to

---

[1] *Journ. Roy. Geog. Soc.,* xxxv. (1865), p. cxlvi.

the Royal Society of your flourishing colony, and see what a
new world is being called into existence by the talent and
energy of our countrymen.   William Merivale has done
good service in smothering that grovelling and unworthy
sentiment of a few *doctrinaires* as to the inutility of our
colonies.   I am furious when I read their cold and heart-
less reasoning, and I shall take good care to show at
our evening affairs how warmly the Australian colonists
support the mother country, and sympathize with it."

It was not therefore the mere spirit of curiosity which
prompted his strong desire to see Australia more thoroughly
explored.   He regarded the vast island as a boundless field
for relieving the over-peopled mother country, and looked
forward to a time when, before the energy of British
colonists, all minor difficulties would disappear, and every
fertile tract of land, every good natural harbour, would be
made available as a starting-point for fresh enterprise in
the endeavour to connect the material riches of Australia
with the wealthy marts of the Eastern hemisphere.[1]   Among
the names of naturalists, politicians, poets, and other celebri-
ties given by the explorers of Australia to the mountains,
rivers, or other features of the land, that of Murchison occurs
in different and far separated regions.   Here it is a wide
county, there a bold mountain range or a broad river, which
perpetuates the name of the geographer who took so active
an interest in the opening up of the country.

African exploration had engaged in a special measure the
attention and fostering care of the Geographical Society.
In every volume of the Journal there had been papers upon

[1] *Op. cit.* p. cxlvii.

that subject, and in later years these papers, embracing the discoveries of Livingstone, Burton, Speke, Grant, Baker, Du Chaillu, Von der Decken, Baikie, and others, had acquired an increasing, and indeed absorbing interest. It was true that in most cases the full and detailed narratives of discovery appeared in the form of volumes of travel, published independently by the explorers themselves. But the first sketch of what had been explored often came to the public through the Society, and when the travellers returned to this country, it was naturally at the Society's meetings that they first presented themselves publicly to their fellow-countrymen. They were sure to meet there with a hearty welcome, and a general appreciation of their courage and endurance in extending the progress of geography.

But it was not merely by empty compliment, or even by medals and votes of thanks that the Geographical Society testified its zeal for exploration. Thanks to the great increase of its membership, it possessed an annual income out of which it could make grants to aid either its own associates, or others who were struggling to enlarge the boundaries of knowledge. Considerable sums of money had in this way at different times been expended to assist discovery in all parts of the globe. But besides what the Society itself contributed, Government had on two occasions, in 1856 and again in 1860, intrusted the Society with the expenditure of sums of £1000 and £2500, to assist in the exploration of East Africa. With the conception and execution of the later journeys of Livingstone, therefore, the Geographical Society was intimately associated, sharing in the honour of having assisted and encouraged the greatest of modern explorers,

In all this work Murchison, whether as President, or as
a Member of Council of the Society, took a leading part.
He sought to make the travellers his personal friends, and in
many helpful ways showed his kindly feelings towards them.
We have already seen how he exerted himself in the organizing of the Livingstone festival. When a long interval
had elapsed without tidings from the traveller on the Zambesi, he sustained the hopes of his countrymen, and refused to
listen to any doubt as to Livingstone's safety. In due time
his friend returned again to this country, after the successful
survey of the Zambesi region. Murchison then proposed to
the Council to send him out once more as leader of a well-
considered expedition to ascertain the true watershed of
Central Africa. In announcing to his brother geographers
the final adoption of this proposal, and the early departure
of the great traveller on this enterprise, the President
sketched the general plan which had been drawn up for
the exploration of the region between the Lakes Nyassa,
Tanganyika, and Victoria Nyanza, and the settlement of that
earliest of geographical problems—the true sources of the
Nile.

This was Livingstone's last expedition. It is still fresh
in the recollection of every one how heartily the President
of the Geographical Society identified himself with the subsequent progress of that prince of explorers. The names of
Livingstone and Murchison came naturally together to men's
lips, and it almost seemed now and then as if people imagined
that the fate of the traveller depended on the heroic firmness
with which his safety was maintained at home by his friend.
It will not soon be forgotten that when, after many months,
no news arrived from the traveller, and when a report arose

of his death, Murchison threw himself with the whole energy
of his nature into the task of convincing his countrymen that
Livingstone was safe, and successfully prosecuting his task.
In the early part of 1867 in particular much public anxiety
existed regarding the fate of the African explorer.   Circum-
stantial intelligence of his death arrived and obtained general belief.   Murchison, however, by speeches at the Geographical Society, and letters to the newspapers, sought in
every way to discredit this report and to maintain his faith
in the skill, good fortune, and admirable constitution of
his friend.   He succeeded at last in inducing the Government to send out a boat-expedition to the head of Lake
Nyassa to investigate the story of the Johanna men.   While
warning his countrymen that even a year might pass without further intelligence, he indulged in a jubilant anticipation of Livingstone's return.   In due time the expedition
sent home satisfactory evidence of the falsehood of the report,
while letters from the missionary himself eventually offered
still more welcome proof of the soundness of Murchison's
judgment in the matter.   But again long months passed
away, bringing no tidings from the traveller, and again
popular rumour began to pass stories of his death.   The
champion at home would listen to none of these, but maintained that he had either struck westwards, following the
supposed drainage of the Lake Tanganyika, and would in
due time appear on the Atlantic coast, or that he had traced
the country northwards from that lake to Baker's Albert
Nyanza, and would first be heard of on his triumphant way
down the Nile, having at last solved the problem of the
sources of that great river.

Looking back upon his scientific career when not far from

its close, Murchison found no part of it which brought more
pleasing recollections than the support which he had given
to African explorers—Speke, Grant, Baker, and notably to
Livingstone. " I rejoice," he said, " in the steadfast pertinacity with which I have upheld my confidence in the
ultimate success of the last named of these brave men. In
fact it was the confidence I placed in the undying vigour of
my dear friend Livingstone, which has sustained me in the
hope that I might live to enjoy the supreme delight of wel-
coming him back to his country." But that consummation
was not to be. He himself was gathered to his rest just
six days before Stanley brought news and relief to the forlorn traveller on the banks of Lake Tanganyika. And
Livingstone, while still in pursuit of his quest, and within
ten months of his death, learned in the heart of Africa the
tidings which he thus chronicled in his journal : " Received
a note from Oswell, written in April last, containing the
sad intelligence of Sir Roderick's departure from among us.
Alas ! alas ! this is the only time in my life I have ever
felt inclined to use the word, and it bespeaks a sore heart.
The best friend I ever had,—true, warm, and abiding—he
loved me more than I deserved : he looks down on me still.
I must feel resigned to the loss by the Divine will; but
still I regret and mourn."

Another proof of its zeal for the exploration and development of the interior of Africa, was furnished by the Society
in 1864, when it offered to embark £1000 in aiding such an
examination of the White Nile as would lead to a commercial intercourse between Egypt and the countries of the
equatorial kings visited by Speke and Grant. That project
had to be suspended chiefly on account of political causes,

But the idea was afterwards carried out in another way, and on a grand scale, by the Egyptian expedition under Sir Samuel Baker.

From an early time the subject of Arctic exploration had possessed a kind of fascination for English sailors. We have already in the course of this narrative watched the departure of Franklin and his companions on their disastrous voyage. The good hope with which Murchison bade them God-speed had slowly died away. In the same spirit with which he in later years espoused the cause of Livingstone he continued to hope when almost everybody else had ceased to do so. In his address of 1857 to the Geographical Society, he returned, but with mournful feelings, to the sad though stirring story of Arctic search. He clung to the hope that still in some less savage inlet a handful perhaps of the lost ones of the 'Erebus' and 'Terror' might be carrying on a precarious existence among friendly Esquimaux. He had failed to induce the Government to renew the search for the traces or records of the missing ships, even though reliable information had been obtained as to where the next efforts could probably be successfully made.[1] He now appealed to his countrymen for their generous sympathy towards Lady Franklin. That noble-hearted woman and devoted wife, undaunted by the refusal of the Government, had herself, with such assistance as her friends could give her, equipped yet another vessel, the 'Fox,' which, under the command of an experienced Arctic navigator, Captain M'Clintock, she sent forth to seek once more for tidings of the fate of her husband and his

[1] The petition to the Ministry was drawn up by him.

comrades. The President of the Geographical Society, warming with his theme, exclaimed :—" May God crown their efforts with success, and may M'Clintock and his companions gather the laurels they so well merit, in their noble endeavour to dissipate the mystery which shrouds the fate of the 'Erebus' and 'Terror' and their crews! If however this last effort, which, in the absence of other aid save that of her friends, Lady Franklin is now making, should fail in rescuing from a dreary existence any one of our countrymen, and should not even a plank of the 'Erebus' and 'Terror' be discovered—still, for her devotion in carrying out the exploration of the unvisited tracts wherein we have every reason to believe the ships were finally encompassed, every British seaman will bless the relict of the great˥ explorer, who has thus striven to honour the memory of her husband and his brave companions.

" My earnest hope is that this expedition of Lady Franklin may afford clear proofs that her husband's party came down with a boat to the mouth of the Back River in the spring of 1850, as reported on Esquimaux evidence by Dr. Rae, and thus demonstrate that which I have contended for, in common with Sir Francis Beaufort, Captain Washington, and some Arctic authorities, that Franklin, who in his previous explorations had trended the American coast from the Back River westward to Barrow Point, was really the discoverer of the north-west passage."[1]

The hope thus generously expressed was duly realized when, in 1859, M'Clintock brought back the records of the lost expedition, which showed that before he had succumbed to his fate, Franklin had really boated from sea to sea, and

[1] *Proc. Roy. Geog. Soc.*, xxvii. (1857), p. cxcvi.

had thus effected that passage which for centuries had been the dream of navigators.

The vast expenditure of life and treasure which Britain had made in quest of the North-west Passage, and the uselessness of that passage when at last discovered, had not by any means extinguished the thirst after Polar adventure. For some time indeed little had been done in Arctic research. In 1867, however, the subject began to receive renewed attention. No one now sought to encourage any further exploration of this commercially impracticable North-west passage. But it was contended that the North Pole might be reached, and that much interesting and important geographical work remained to be done in that region. The desirability of renewed exploration had been pressed upon the Government by the British Association and the Geographical Society, but without success. The geographical authorities were far from being agreed as to the proper avenue by which the attempt to reach the Pole should be made. Four routes had been advocated. One by Spitzbergen, one by the north-east coast of Greenland, one by Behring's Straits, recommended by French geographers, and one by Smith's Sound, which was warmly espoused in England. This want of agreement among those who had given special attention to the subject seemed to furnish a ready justification of the unwillingness of the Admiralty to re-engage, in the meantime, in a task where already so many precious lives and so much valuable property had been lost. Murchison, in successive addresses to the Geographical Society, dwelt emphatically on the importance of taking action in this matter, showing what Swedes and Germans had done, and how much might now be accomplished with the increased knowledge and improved appliances which could be brought to bear upon the question. He took occasion, likewise, to point out the desirability of further Antarctic exploration, especially with a view to fixing stations for observing the transit of Venus in the year 1874.

How these geographical efforts appeared in the President's correspondence, may be illustrated by a letter to Sir William Denison :—

" 11*th June* 1865.—My dear Sir William,—I am ashamed of my laxity as a correspondent, but truly if you lived in this world of London, and were in as many whirlpools of employment as I am, you would admit that, with the best intentions, I cannot do justice to my own desire to endeavour to reciprocate in some measure the kindness you show to me, and the interesting communications you send to me. Now that I have a Sunday evening free (and having already written a dozen of notes), I find time (my anniversary Address having gone finally to press) to write a few lines, which, after all, may be of very little interest to you.

" As I really have given the Geographical Society a considerable impulse, the good Fellows over whom I preside recompense me with warm thanks, and then express their desire to work the old mill-horse to death, so I am in for another year—all rules being broken. The past year has been productive of good discoveries, so I took it upon me to review the progress of Geography since the formation of the Society, and to point out the enormous amount of the desiderata to be accomplished by ourselves and our successors. But you will hear all this very soon, as my Address will be printed off in a week.

" The chief feature of our efforts has been to get up a

North Polar Survey. Independently of the great geographical problem to be solved, the fact is that our navy
lacks something to inspirit them. The science of the
navy has been going down for years, and they think of
nothing but how to protect themselves in iron-bound vessels.
But here again I must refer you to all our discussions and
to my Address for the warm support we have received from
the Imperial Academies of St. Petersburg and their excellent President Admiral Lütke, himself an Arctic voyager. I
think I should succeed if I had only the Duke of Somerset
to deal with. But alas! his first lord and enlightened
secretary are dead against us and every scientific Society of
Britain and the Continent.

"Oh for the good days of adventure, of Raleigh and
Drake, and when Hudson and Baffin made voyages of discovery in sailing vessels of 70 or 80 tons, which our de-
generate *Lords* shrink from attempting with all their means
and appliances. I could not help letting off in the conclusion of my Address, and saying that a more enlightened
posterity would applaud our efforts, though they are now
opposed by a dull mass who look only to profit and loss.

"Livingstone is about to be off on his third tramp in
Southern Africa, to determine, if possible, the true watershed
of the region north of his own great lake Nyassa.

"Sherard Osborne, to my great regret, has left as a Bom-
bay Engineer, but his heart is still with us, and even now
he writes to me warmly commending a project of Captain
Allen Young for the opening out of the Korea to British
exploration and commerce. With this and the excellent
service of Colonel Pelly, who has determined the real position of places in Arabia recently visited by Gifford Pal-

grave, we close our Geographical Session. The great folks also close their political session, and we are about to expire. I will get away to some place on the Continent during the turmoil of an election in which I take no part.[1]   But in September I must again be at my post, to preside over the geologists of the British Association at Birmingham.—Yours sincerely and obliged,   ROD. I. MURCHISON."

In these final efforts of his life, though not immediately successful, Murchison deserves the praise of having clearly foreseen that the British Government could not much longer refuse again to co-operate in the great work of Polar discovery. He did not live to witness the sailing of the 'Challenger' to extend our knowledge of the depths of the seas, even as far as the Antarctic ice-barrier, or the preparations for the Arctic expedition of 1875 in search of a passage to the North Pole. Had he survived so long he would have seen reason to modify his opinion that the love of adventure had died out of the British navy, for his friends at the Admiralty could then have assured him that the vessels selected for the North Polar quest might have been manned with officers, so large was the number who volunteered for that service.

In fine, probably no better summary of Murchison's

[1] Though not a politician, and, as appears from this passage, abstaining from interfering in Parliamentary contests, Murchison at times could break through this habit and become for the nonce a keen partisan. I remember being with him on the afternoon of the day when Mr. J. S. Mill lost his seat for Westminster. I was at once eagerly questioned by my friend as to the last news from the polling-booths, and on my remarking that Mr. Mill bade fair to be defeated, he rubbed his hands with great glee, saying, "I was out by eight o'clock this morning to vote against him. I would walk the shoes off my old feet to have the fellow turned out after his infamous conduct towards Governor Eyre in the matter of the Jamaica insurrection."

services to Geography in connexion with the Geographical Society could be given than in the words of Sir Bartle Frere, who succeeded him in the Presidential Chair, and who, in conveying the medal awarded to him by the Council, spoke of him thus :—" It is no exaggeration to say that during the past thirty years no geographical expedition of any consequence has been undertaken in our own or, I believe I might say, in any other country, without some previous reference to him for advice and suggestion, often entailing laborious research and correspondence."[1]

In passing from the Geographical Society we may take notice of one feature of the anniversary Addresses on which Murchison always laid great stress—the obituaries of deceased members. These afforded an opportunity, of which he never failed to avail himself, to sketch the services and good qualities of old scientific friends and companions. Most of his compeers in the Geological Society were likewise enrolled among the geographers. Hence year by year he had occasion to pronounce an *éloge* over the grave of one after another of the early magnates in geology. At one time it is the genial Buckland to whose memory he has to pay a kindly tribute, remembering not only the lasting services of that able man to science, but the many kindnesses which he had himself received, and not least among these, the friendly guidance which led him to the banks of the Wye in 1831, and indirectly to the Silurian System and his after fame. Again he has to chronicle the quenching of another of the lights under which geology in its early days spread and prospered in England— William Conybeare. Of the leaders who upheld the science

[1] *Proceedings of the Geographical Society*, 1871, p. cxxxvi.

when he first began to study it, the author of the Silurian System was indebted to no one more deeply than to this able observer and admirable writer. Conybeare and Phillips' *Geology of England and Wales* had been, as he said, his scientific Bible. From his earliest geological paper onwards, the influence of that book may be traced in all his geological writings. These obligations he gratefully acknowledged. At another anniversary, when death had been busy among the leaders of science, and especially among the President's own circle of friends, he had to record the loss of Robert Brown, to whom he was sincerely attached; Alexander von Humboldt, from whom he had received so many proofs of respect and esteem, and to whose assistance and stimulation he now gratefully recounted his obligations; Hallam, one of the most welcome guests at his gatherings of scientific, literary, and artistic friends; the Archduke John of Austria, the frank, open Styrian prince, with whom he had been so delighted among the valleys of the eastern Alps.

CHAPTER XXVIII.

THE LAST GEOLOGICAL TOUCHES.

DURING that closing period of Murchison's life which embraced the years subsequent to his appointment to the Geological Survey, and of the events of which some account has been given in the previous five chapters, geology had evidently no longer undisturbed sway over his thoughts and actions. Nevertheless, amid the increasing exactions of official life, and in spite of the ever-growing demands of the Geographical Society, he found opportunity every now and then to strike back again into the geological pathway, and to link himself with the onward progress of his younger contemporaries. It was during this period that he achieved his success in working out the problem of the Highland geology. During this time, too, he made those excursions in Bohemia and elsewhere on the Continent, as well as in the north of England, which enabled him to place the Permian rocks of Germany and Britain in a more satisfactory arrangement and parallelism. But besides these contributions, which arose directly out of field-work and gave the results of his own observations on the ground, there were others which dealt not so much with his own labours as

with those of his friends and fellow-workers in geology—
digests of the progress of research among the palæozoic
rocks, criticisms of contemporary opinion regarding questions of theoretical import, and more or less vigorous pro-
tests against the spread of what he considered to be rank
heresy from the orthodox geological creed.

Though many of these writings may have no special
importance in themselves, they were the last efforts of a
man who has left his mark deeply upon the literature of
geology, and as such seem worthy of notice here. It is proposed therefore to devote this chapter to a brief notice of
them, placing them in such lights as may best reflect the
character of the man, and connecting them at the same time
with the general onward march of the science.

Of all the pieces of scientific work now to be referred to,
the successive editions of *Siluria* deserve the first place.
We have already seen how that volume was at first
elaborated, partly in the field among the rocks, and partly in
the library in the midst of memoirs and notices on palæozoic
rocks contributed by geologists all over the world. The
subsequent editions involved labour of the same kind.
A few extracts from the geologist's letters will best illustrate the desultory way in which this literary work was
carried on :—

"*February* 21, 1857.—MY DEAR M. BARRANDE,—On the
20th of this month I received for you, at the anniversary
meeting of the Geological Society, the Wollaston medal,
which had been adjudicated to you by the Council of that
body, for your works on the Silurian basin of Bohemia, which
justly raised you to so eminent a place among the geologists
and palæontologists of Europe, and have (as I wrote down

the words which will be printed) ' won for you the admiration of your contemporaries.' . . .

"I wish I could report to you that I had made much progress in the second edition of *Siluria*, but my public business, and the numerous calls on my time for transacting the new duties of President of the Royal Geographical Society (now one of the most popular Societies in the metropolis), have sadly retarded me, and a few chapters only are printed. These early chapters, however, are the difficulties, for they are much improved and altered. In the first place, by revisiting my old typical region of Siluria, I have strengthened my base so essentially, that it stands firmer than ever. . . .

"The next important new feature is the separation in North and South Wales of the Llandeilo from an overlying formation, which is my original Caradoc. . . .

"In order to remove all ambiguity, I have placed the zones with *Pentameri*, of various species, and *Atrypa hemisphærica*, as of intermediate character, terming the strata 'Llandovery Rocks,' the lower half of which is intimately connected with the true Lower Silurian by numerous fossil species, and the uppermost or thinner part, containing *Pentamerus oblongus*, with the Upper Silurian through certain species.[1]

"It has transpired that most of the rocks mapped by the

[1] Originally, as we have seen, Murchison had no breaks in any part of his Silurian series, from the base of the Llandeilo to the top of the Ludlow rocks. It was sadly against his inclination that he was driven to admit that, in this respect, his series was not the unbroken whole which he had represented it to be. Sedgwick was the first to point out the great unconformability at the base of the Upper Silurian ; and the Geological Survey, especially Mr. Ramsay, its director, and Mr. Aveline, afterwards traced other breaks.

Government surveyors as Upper Caradoc, in North Wales, are rather the base of the Wenlock formation in a sandy or gritty state with associated slates, but these are local varieties."[1]

"Poor Dumont has passed away too soon! He was a clear-sighted field geologist, and an admirable lithologist and mineralogist. If he had been earlier imbued with a true veneration for organic remains, he would have been one of the first men of our age, or rather *nulli secundus.* The Belgians know how to honour their eminent countrymen. When will any geologist of Britain be put into the grave with so much pomp and circumstance as our departed friend—three battalions, with archbishops, senators, professors, scholars, all vying with each other?"

To Professor Ramsay he writes :—

"I have had a long letter from Oldham, thanking me twice over for my warm intercession with Lord Canning, which has, he says, been the cause of the grand augmentation and indeed of the real setting up of the Geological Survey of India. They are now endowed, and are to have library, laboratory-rooms, and a museum—in short, a Jermyn Street: I would add, and with the unspeakable advantage of being unfettered by ' My Lords'. . . .

"I am like you, ' *totus in illis,*' and looking over a beautiful garden, whilst I am re-writing and improving the first chapters of *Siluria.*"

But the absorption referred to in the last sentence suffered

[1] He contended that these breaks were local, because they did not occur in other countries. But that between Lower and Upper Silurian is not a mere local or British phenomenon. The Geological Survey at first accepted his classification, and put the Pentamerus beds in the Caradoc series. The error had been corrected some years before the date of the above letter.

many interruptions. Take one sample of the "*totus in illis.*" —"At night went to the annual ball at East Grinstead. Danced a quadrille with Mrs. Mortimer West, and a tempête with the youngest Miss Stirling, Lady Caroline's lovely daughter, and a reel with Lady Arabella. Pretty well for a chicken of 1792, who had been geologizing all day. Came back in a storm to Buckhurst."

Here is his report at a later time to his old friend Sir Philip Egerton :—"*Siluria* will be out in a fortnight, with nearly 200 new figures of species in woodcuts, new tables, coloured lithograph of Assynt, and all sorts of novelties, and much additional matter.

"On Wednesday next I throw off No. 2 of my N. Highland and Orcadian contributions, and on 15th December the Old Red of Elgin comes off, followed by Huxley's description of the wonderful reptile *Stagonolepis Robertsoni*, which must have been many feet in length, and had a swinging tail. The singular big cast which I brought away in my portmanteau and showed you at Leeds, which I told old Duff seemed to me vertebral, has, after being talked of as a Cephalopod, been proved to be the tail of *Stagonolepis*. We have lots of his footprints. Huxley makes him a Thecodont.

"I am getting a charter for the Geographers, and going to Lord Derby for apartments to contain both the Geological and Geographical Societies at Burlington House, so as to constitute there an Institute of Science."

When at last the edition of *Siluria* referred to in the foregoing extract was published at the beginning of 1859 a copy was duly sent to Sedgwick. That old comrade had expressed himself very strongly, in his *Synopsis of Palæozoic Fossils*, regarding the standing subject of controversy. Since the appearance, in 1855, of the strongly worded, but somewhat tedious, introduction to that publication, in spite of Murchison's efforts, it was impossible to conceal the fact that the difference remained no longer a mere scientific one, but had descended, as such differences almost inevitably do at last, into personal estrangement. He got no acknowledgment of the book, nor of the letter written " in his old fashion," which was judged by himself and a common friend to be best. After waiting a week he sent a note to the Master of Trinity, with the half-hinted hope that the latter might be able to send some good news, yet with the expressed fear that Sedgwick's silence was intentional, and would be lasting. The letter to the Woodwardian Professor was as follows :—

" *Jany.* 20*th*, 1859.—MY DEAR SEDGWICK,—I have sent a copy of my new edition of *Siluria* for your acceptance, earnestly hoping that the passages relating to yourself in the Preface, p. viii, and the alteration of a phrase or two in the body of the work, may remove from your mind the impression produced by the perusal of the first edition.

" Time rolls on, and as we passed many a happy day together, I trust that you will have some gratification in turning to these pages, particularly those relating to the Highlands of Scotland. Little did we think, when we first united the yellow and white sandstones of Elgin with the Old Red, that those beds would be found to contain such reptiles as the *Telerpeton* and *Stagonolepis!*

" Clinging to the hope that the only bitter sorrow I have experienced in my scientific life may pass away, and that your old friendly feelings towards me may return, I am always, my dear Sedgwick, yours most faithfully,

" ROD. I. MURCHISON."

A reply to this letter was written from Dent, and beginning with " Dear Sir Roderick," contained a mere acceptance of the gift (which was still lying at Cambridge), without an allusion to the hope of pacification. It was afterwards tied up with some other correspondence, and marked on the back "Sedgwick's last letter. How different from the 'Dear Murchison' of former days !"

Two years afterwards Murchison renewed the attempt to re-open friendly intercourse with his former comrade. In the May of 1861 he received the degree of LL.D. from Cambridge. Many a pleasant and many a sad memory were awakened by the visit. Some of these found expression in the following letter :—

" *The Lodge, Trinity College,* 22*d May* 1861.—DEAR SEDGWICK,—I cannot be once more in Trinity College without having brought vividly to my mind our former friendship and your kindliness to me on many an occasion. Permit me to assure you that I have at this moment as strong a regard for you as ever, and that although I yesterday received an honour which you had secured for me at the installation of Prince Albert, if the University law would have permitted, I consider this, and all worldly distinctions that might be bestowed on me, as nothing, in my estimation, compared with one kind letter from yourself, in your old manner. I should, of course, have infinitely preferred to make the appeal to yourself in person, and was much disappointed when I found you were not to be present at the Convocation of yesterday.

" After the ceremony I went to visit your museum, and was perfectly astonished at the immense improvements and additions which have been made in and to it since I was in

Cambridge.   I recollect full well how slightingly you always
spoke of the Upper Greensand of Cambridge, and how, when
we were together in Westphalia, you looked to me as a sort
of authority on the 'Malm Rock.'   But who could ever have
imagined that your little Cambridge zone would have given
forth such riches as it has afforded?   I was really quite
astounded at the quantity, variety, and value of the
fossils. . . .

   " Among the few old friends left here I was vexed to see
Hopkins so much broken, though he officiated in the ceremony, and will, I trust, now recover his wonted vigour.   If
you will, at your leisure, gratify me with a line, and let me
know that you are in better health than when I saw you last
at the Athenæum, you will sincerely gratify me.   The days
of 'auld lang syne' are perpetually recurring to me, and
when I transmit to you, as I shall when I go to town, a copy
of my last memoirs on the Highlands, including one to refute
Sharpe's errors about the cleavage of the Highland rocks, you
will see how I recur to your dicta on that point, as well as
to our original observations in Scotland.—Yours always
most truly,   ROD. I. MURCHISON."

   Later in the same year the British Association met at
Manchester, with Murchison in the chair of Section C.   He
gave an opening address, the feeble character of which is probably accounted for by the following entry in his journal :—
" I am much out of health, and certainly all the worse for the
coolness and chill with which my former friend Sedgwick
received every effort I made to be on our old terms with him.
I could not have believed that a man whom I had always
looked on as the most hearty and generous of beings should

have turned upon me with unforgiving tenacity, because, as he said, 'I had gathered all the cream of our churning in the rocks of Wales.' . . . I wish to avoid, as far as possible, another public demonstration like that which took place at Manchester, and which really vexed me beyond measure."

Sedgwick, stung to the quick by what he thought to have been unfair treatment on the part of his old companion Murchison, for which he called in vain for apology and retractation, withdrew from all further intercourse with him and from the Geological Society. On two occasions only did he afterwards write to his friend, the last time being when the cords of sympathy and affection were loosed within him at the sight of that friend bowed down with sorrow by the death of his wife.

The last edition of *Siluria* was published in the autumn of 1867. If we compare it with the first edition of the work we see how much progress had been made in palæozoic geology during the interval of thirteen years. Among other changes, the Laurentian system had been established. The Cambrian, which had been purposely omitted from special description in the first edition, now receives recognition in the limited sense in which the term was used by the Geological Survey; the physical break between Lower and Upper Silurian contended for by Sedgwick is now admitted, and a zone of passage between the two series is described under the name of " Llandovery rocks." The important discovery that the rocks of the Scottish Highlands, formerly thought to be of an origin anterior to the existence of life upon the globe, consisted really of altered Lower Silurian rocks, is pointedly dwelt upon. Great improvements and additions are introduced into those portions of the book treating of organic remains. Especial attention is likewise given to the foreign equivalents of the British formations, and thus the book is made truly an indispensable handbook to all geologists who are intent upon deciphering the history of the oldest fossiliferous rocks. No better evidence of the practical utility of the work could have been given than the fact that, in spite of its technical character, its total want of literary attractiveness, and its high price, four editions were called for in thirteen years.

Quitting his own special domain of palæozoic geology, Murchison in his later years took considerable pains to reiterate his faith in a former greater intensity of Nature's operations, and to oppose the doctrines of the opposite or uniformitarian school. His opinions on this subject are strongly expressed in the closing pages of the last edition of *Siluria.* He used often to announce them from the chair of the Geographical Society, but the greatest vigour of language was reserved for his private correspondence. Some illustrations of his opposition may be given here.

For some years there had been growing among the younger and more active geologists of Britain a conviction that the old doctrine of Hutton as to the origin of valleys by the erosive action of running water—a doctrine which, in spite of the admirable confirmation of it adduced by Mr. Scrope from Central France, had never been generally adopted—was substantially true. Foremost among those who maintained this view, and enforced it by cogent argument and illustration, were the Directors of the Geological Surveys of Great Britain and Ireland—Ramsay and Jukes. The former, moreover, started and worked out the remarkable idea that besides the excavation of valleys by river-action and the slow washing of the land by rain, there has been an extensive erosion of hollows and basins by glacier-ice, and that to this process we must attribute the great predominance of rock-basin lakes scattered over those tracts of the northern hemisphere which can be shown to have been buried under land-ice. As may be supposed, the President of the Geographical Society regarded these doctrines as rank heresy, not to say sheer nonsense. He opposed them chiefly in his addresses to that Society; but his opposition, though vigorous enough in its language, dealt more in strong denial and protest, with the citation of the crowd of geological authorities who sided with him, than in serious argument.[1] The force of evidence had constrained him to yield somewhat of the old exclusiveness with which he had fought for his icebergs, but having given up some points, and consented to admit the power of glaciers to polish and score the face of a country, and to pile up huge moraine-mounds, he felt himself free to set his foot down firmly and refuse to go a step further in the way of excavation, as his friends, the "ice-men," would have had him.

In science, as in all other matters of human thought and progress, it is not given to many men to retain complete possession of their youthfulness and pliability of mind up to the end of a long life. The ardent spirits of one generation are apt to become in the next what their successors

---

[1] In one of these Addresses he gave great prominence to the question of glacier-erosion, especially in reference to Professor Ramsay's then recently broached doctrine. Not content with the publicity given by the Journal of the Geographical Society, he extracted that portion of his Address, and circulated it far and wide among the geologists of Europe and America. Professor Ramsay's reply in the *Philosophical Magazine* (October 1864) was valuable as a protest against the attempt or tendency to crush opposition by weight of authority.

irreverently term "fogies."   When Greenough stood out
stoutly for the integrity of "Old Greywacke," and refused to
countenance any of the New Lights, Murchison, as we have
seen, enjoyed some quiet fun at his expense, and laughed at
his stubborn adherence to the "antiquas vias."   But the
whirligig of time brings in his revenges.   It was now
Murchison's own turn to protest vehemently against the
spread of notions which, nevertheless, in spite of his opposition, were every year steadily gaining ground.   So much
did he take to heart the backsliding of the younger school
from what he thought the true faith, that besides taking the
pains to circulate his protest among scientific men all over
the world, he could hardly refrain in his letters to old friends
from finding some excuse to introduce the subject.   The
subjoined letter to Sir William Denison (6th October 1864)
may serve as an instance :—

" In my anniversary Address to the Geographical Society
you would see the pains I have taken to moderate the ice-
men, who would excavate all the rock basins by glaciers
eating their way into solid rocks.   I have printed this
excerpt from my Address in a separate pamphlet, headed ' On
the Relative Powers of Glaciers and Floating Icebergs,' and
this, as well as my Bath oration,[1] goes to you.

" In seconding the motion of thanks to Lyell for his
address at Bath, I felt bound to say a few words in defence
of my opinions as to the grander intensity of causes in old
geological times than in the present or Man period ; and as
Lyell had used the words ' some great convulsion and fracture,' to account for the great rent and fault out of which

---

[1] That is, his Address to the Geographical Section of the British Association.

the hot Bath water flows, I said I was happy to receive that indication of the right view, and that I should in future range my friend Sir Charles along with myself among the ' convulsionists.'

" And again, I entirely disagree with him when he adverts with triumph to the discovery of animal life in the old Laurentian rocks of North America, that this is any indication that we have here ' no trace of a beginning.' On the contrary, the only animal which has been found, being a zoophyte, adds nothing and changes nothing in the general argument founded on the indisputable facts recorded all over the world, viz., that there has been a progression of creation from the lowest grades of animal life up to man.

" It is mere special pleading to sound a trumpet of glorification over the demolition of the word ' Azoic.' Whenever I used that word I specially guarded it by saying that although as yet nothing like animal life had been found in the very lowest known rocks, yet the discovery of Fucoids and Zoophytes might be found, and if so would not alter one jot the order of successive and rising creations.

" As to your view concerning the desiccation of countries by the destruction of forests I entirely agree with you, and on this head Colenso is certainly weak and wrong. Nothing can be more demonstrative to that effect than Cyril Graham's description of the numerous ruined cities close to each other in the Hauran, where now there is not a drop of water, and nothing to sustain human life. The same phenomena is now being worked out on a gigantic scale in the great and recently highly cultivated and productive region east of Palestine, as vouched for by the Rev. Mr. Tristram, who has just returned from these parts. He gave

us a very eloquent description of all the natural phenomena
in the valley of the Jordan, and has completely swept away
the delusion under which we all laboured, through the inac-
curate accounts of travellers, that the great depression was on
a line of volcanic eruption and subsidence. On the contrary,
both sides of the valley consist of regularly stratified limestone dipping into the valley, and forming a simple trough.

" I wish we had had a few of your 57 inches of rain in
England. For in the last century there was not such a want
of water, and even now the east wind is steady, so we want
you as ' Jupiter pluvius ' to come among us.

" I have always foreseen great embarrassment and difficulties without end in the satisfactory arrangement of the
governments and peoples of Italy under one head, and now
the play is beginning. Well may the brave Piedmontese
be crestfallen and indignant when they see that, after losing
Savoy and the key to Italy in the Alps, they are doubly
taxed and Turin is no longer a capital. The Tuscans, who
never would or could fight—so all the old French officers
have told me—are for a while to be flattered into the idea
of their city being the capital. But this is all fudge,
for in reality the people who cry ' Italia una ' wish for
Rome. Then again the Neapolitans are full of discontent
at not having their city now what it was for so many
centuries, and the Pope will blow up many a conflagration
before he is driven out of the Eternal City. I have lived
much in Italy and Sicily, and, though I should be very
happy to see the dream of poets and the English press
realized, I feel convinced that the elements of discord so
abound that unity will not last now any more than in ages
gone by.—Ever yours sincerely,   ROD. I. MURCHISON."

In no part of his career did Murchison show his shrewd common-sense more than in the way he usually avoided any public expression of his opinion on questions in science which he had not specially studied, or which lay wholly outside of his own circle of inquiry. Few, indeed, can exactly gauge their own capacity and qualifications so as to know exactly how far they may safely advance in any given direction. Murchison's natural caution, however, made his slips of this kind comparatively few. For example, he entertained a very strong repugnance to the views promulgated in the great work of Darwin on the *Origin of Species*, and he believed firmly that the geological evidence of the older formations lent no support to these views. But he took care not to proclaim his opposition from the house-tops, and indeed honestly confessed himself not qualified by knowledge and experience to discuss biological subjects. His sentiments, however, used to be pretty freely expressed in conversation and in letters to his friends. A few extracts will show the kind of argumentation which he thought so formidable to the Evolutionists.

To Professor Harkness he says :—" If you read the work of Darwin on the *Origin of Species*, which has given us an earthquake shock, you will easily see that in reality my geological postulates, if not upset, destroy his whole theory. He will have no creation—no signs of a beginning—millions of living things before the lowest Silurian—no succession of creatures from lower to higher, but a mere transmutation from a monad to a man. His assumption of the position of the Lyellian theory, that causation never was more intense than it is now, and that former great disruptions (faults) were all removed by the denudation of ages, is so gratuitous, and so entirely antagonistic to my creed, that I deny all his inductions, and am still as firm a believer as ever that a monkey and a man are distinct species, and not connected by any links,—*i.e.* are distinct creations.   The believers in a lower, and a lower still, have never answered, and cannot answer, the fact that the rich marine Lower Silurian fauna is invertebrate, and that the Cambrian rocks of Ireland, Wales, Shropshire, and the north-west of Scotland, though less altered than the Lower Silurian, have afforded nothing distinct which is higher than an Oldhamia or a worm."

Again, in a more forcible style, he remarks to Sir William Denison :—" I am a geologist of the school of Buckland, Sedgwick, De la Beche, Greenough, and I may add, of myself.   I flatter myself that I have seen as much of nature in her old moods as any living man, and I fearlessly say that our geological record does not afford one scintilla of evidence to support Darwin's theory."

" Recently we have had the grandest trumpeting about the discovery of the *Eozoon Canadense* in the Laurentian rocks below all Cambrian.   And what does it amount to ? Why, simply that the lowest imaginable order of zoophyte is found in the lowest discoverable rock.   It changes nothing. We are just where we were.   Simply the lowest created things are found in a stage lower.   This only confirms the doctrine of a commencement with the lowest grades of creation, and a succession in after ages to higher and higher types of life successively.   As for the transmutation of types, I look upon it as simply an ingenious piece of sceptical puzzling without the least basis."

The curious confusion of thought here, regarding the bearing of the grade of the older palæozoic fauna upon the doctrine of evolution, re-appears in other letters.   It would seem as if the Silurian geologist had been so long battling against the favourite uniformitarian doctrines of Sir Charles Lyell, in the organic as well as the inorganic history of the world, that he could not quite realize the complete change of position which the author of the *Principles of Geology* took up after the new era of thought marked by and consequent upon the appearance of Darwin's work.

The publication of Lyell's *Antiquity of Man* drew forth in the same way these private but strong expressions of dissent from that part of the work which treated of the origin of species. In regard to the question of the antiquity of the human race Murchison had no strong prejudice, and was willing to accept the new doctrines as established on fair evidence. To the author of the work he acknowledged as follows the receipt of the presentation copy :—

" I have to thank you for your very acceptable present of your new work on the *Antiquity of Man.* Who would have thought when we began to write as geologists that any one could have in our lifetime reached such conclusions as those which you have deduced ? You have indeed taken a fine bold course and have opened out a grand and broad field of inquiry. I am thankful for a slight move on your part towards occasional paroxysms, and as we all admit slow accumulations as well as imperceptible elevations we may be all brought into harmony. There are of course some points on which I still hold to my old opinions, particularly as to the impossibility of referring all the water-worn débris of the low country of the Pays de Vaud to the ancient glacier which came out of the narrow gorge of the Rhone. But these have little to do with your main argument."

It might not be supposed from this note that its writer

had anything to complain of personally in the contents of
the volume he had received. But in writing to another
friend he lets out the secret of his heart : " I presume you
will get Lyell's new volume on the *Antiquity of Man,* and
will marvel at and perhaps admire the bold efforts now
made to throw back the origin of our noble species *Homo*
to the accumulations immediately succeeding the glacial
period, when half of modern Europe was either under snow
and ice or icy seas. Huxley's ' Place of Man in Nature '
completes the view by showing us that man is only the
front-rank leader of a succession of apes. This little book
is beautifully written. My gifted colleague runs far ahead
of my knowledge. I must apply to myself ' Ne sutor ultra
crepidam ; ' as yet however I am not a Darwinian, and see
numberless objections to his theory.

" The geological part of Lyell's book is very good, but
has little or nothing new in it. . . . The data collected
by others and reasoned on by them as to flint implements,
etc., were all well known to us. . . . As to myself, I am
only once mentioned, in order to be knocked down *in re*
ice and drift. But having been so mentioned I think my
old friend might have cited my chapters on erratic blocks,
and particularly from my big book on Russia, in the map
of which I laid down for the first time the exact lines
or limits of the eccentric dispersion of the Scandinavian
and Lappish rocks over Russia, Germany, the Low Countries, and Britain."

In a later letter to the illustrious author of the *Antiquity of Man* Murchison bluntly complains of the non-
recognition of his own labours in the department of qua-
ternary geology, and laments the mistake of writing " big
books " like his *Silurian System* and *Russia in Europe,* all

his elaborate details on that subject seeming to have been quietly buried out of sight in those ponderous tombs.

The irritation which the author of the *Silurian System* felt at any proposed alteration of his classification of the geological formations, and the efforts he made to adopt the inevitable changes with as little damage as possible to the original nomenclature, have already appeared in the course of the preceding chapters. A proposal, renewed in Germany, to adopt the term Dyas instead of Permian,[1] was in itself rather of a retrograde character, as it rested upon mere mineral grouping, which, after all, was only local. The progress of geology had abundantly shown that names founded only on the character and arrangement of sedimentary rocks were apt to become singularly inappropriate when applied to the equivalents of the same rocks in other regions. Murchison's protest against the projected change appeared duly in print; but the warmth of his indignation could, of course, only find vent in talk or correspondence with his friends. Take the following as an example :—

"MY DEAR HARKNESS,—My good friend Dr. Geinitz, has, I regret to say, revived the ridiculous term of Dyas as a substitute for Permian. Taking it from Marcou, who had most absurdly applied it to an union of the Permian with the Trias as 'the hard red series,' Geinitz, wholly disapproving that absurdest of projects (for he knows the one is palæozoic, the other mesozoic), still proposes the name as a substitute for Permian, because, forsooth, in the limited tracts of Germany which he knows, the Rothliegende and the Zechstein constitute the group. I have already written a

---

[1] The name "Dyas" was first proposed in 1859 by M. Marcou, and objected to then by Murchison (*Amer. Journ. Science and Arts*, xxviii. 256). His protest against the renewal of it in 1861 appeared in the *Geologist*, v. p. 4.

paper, which I am sending to the printer, to quash this nonsense (for it is truly such) at once.   Trias is the worst word
we ever had in use, for in England it is, as you know, a
Dyas, in other tracts abroad a Monyas, and in others a
Tetraias.   So is it with the Permian.   That name involves
no necessary number of physical or lithological divisions ;
and, proposed by me twenty years ago, it has been in current
use everywhere for fifteen years.

"In Russia it is one great series of alternating limestones, marls, sandstones, gypsums, copper ores, and conglomerates, the Zechstein fossils occurring at various
horizons.   But I need not dilate.   You will read my paper
soon, for I will print it myself, and send it all over Europe
and America. . . .

"The logic, or want of all logic, on the part of good
Geinitz is lamentable.   But the Germans are reviving old
Grauwacke.   Pray answer this by letter to London.—Yours
sincerely,   Rod. I. Murchison."

It was not always, however, from abroad that opposition
to the views of the Director-General arose.   In matters of
theoretical import, such as the origin of valleys and lake-
basins, there could hardly be any violent collision with him,
for though he held tolerably strong convictions on such
questions, they belonged to a branch of geology which he
had not made specially his own ; so that, as he said himself,
his views might be set aside without impugning the value
of those positive data in geology which he had done so
much to establish.[1]   But he could not be expected to remain quite so unconcerned when some of those very posi-

[1] See *Siluria*, 4th Edit., p. 506.

tive data on which he most prided himself were attacked.
Hence, when in 1867 his own colleague, the late Mr. J. B.
Jukes, Director of the Geological Survey of Ireland, began
to throw doubts on the received classification of the
Devonian rocks, he watched the progress of the discussion
with keen interest, but, partly from a wish not to let open
disagreement appear too strongly between himself and his
officers, kept as well out of it himself as might be. Mr.
Jukes contended, in direct opposition to Murchison's long
accepted conclusion, that the Old Red Sandstone could not
be classed with the Devonian slates and limestones without
a violation of all the rules and logic which govern our geological nomenclature. He maintained that the Old Red
Sandstone was something altogether different from and
older than the Devonian rocks, and that the latter must be
ranged with the great masses of slate and grit which underlie the Mountain Limestone in Ireland, and form the base
of the Carboniferous system.

It may be imagined how much Murchison would take
this proposed reform to heart. Honestly and firmly believing that his colleague was making an egregious blunder
in venting such a heterodox notion, he urged him very
warmly to recede from the untenable position. In letters
to other friends he spoke in a compassionate way of the
wilful waywardness of his subordinate, who seemed to him
bent on becoming a geological *felo de se*.

But the bold and skilful Director of the Irish Geological
Survey knew what he was doing. He could not believe,
nor even understand, that any one could possibly be offended
by an attempt on his part to correct what his experience
had taught him to be error. And so, holding clearly to his

purpose, and battling stoutly for the change which he proposed to make, he soon after sank under the insidious and
rapidly advancing malady which carried him off.

It is not necesssary for the purposes of the present biography to enter into the details of this new Devonian battle,
for during his lifetime Jukes had in this question few allies;
and Murchison, who survived him, did not imagine that the
contest would ever again be seriously renewed. But they
who have given most attention to this part of geology will
probably most readily admit that, whether in the way of
contest or not, the question must be re-opened, that the
accepted classification of the rocks of Devonshire is far
from being satisfactory, and that Jukes did a great service
by boldly attacking it and bringing to bear upon it all his
long experience in the south of Ireland, which gave him an
advantage possessed at the time by hardly any one else.

Among the deaths by which the ranks of the geologists
of Britain have in recent years been thinned, none came as
a heavier blow than that of J. Beete Jukes. In the prime
of his life, with no common bodily vigour, with a capacity
for field-geology second to that of no one in this country,
and with a ready power of felicitous exposition, he seemed
to be a man on whom much of the future progress of the
science among us would depend. But it was not only, or
perhaps even mainly, for his rare scientific qualities that his
death was mourned. His qualities of heart were rarer still.
A more joyous, generous, kindly spirit lived not among us.
In the heartiness and hilarity with which he threw himself
into whatever he had to do, he preserved almost the freshness of boyhood. Before the appearance of his fatal para-
lytic affection there could be no lack of solid talk and

J. B. JUKES, F.R.S.
From a Photograph.

playful sallies in any company where his broad burly frame appeared, and his own deep ringing laugh would soon be the loudest there. But with this playfulness there was ever the earnestness of a man who knew that he had a work to do in this world which required of him his whole energy and thought. He had no tolerance for pretence of any kind, showing his impatience sometimes in a way which led those who did not know him to form most mistaken judgments regarding him. But they who enjoyed his acquaintance, and felt the frank impulsiveness and downright honesty of all that he did and said, will for many a year lament his loss, for they never found in his craft a better workman nor in their own circles a truer friend.

This rather desultory retrospect of the geological topics which occupied Murchison's time and thoughts in his last years, may perhaps suffice to bring before the reader their somewhat miscellaneous character, as well as the tenacity with which his opinions regarding them were held by the geologist, and the vigorous style of defence and protest which he used on their behalf. It may now be appropriately closed by a reference to some of the later honours of his life. On the 19th of February 1864 the Geological Society awarded to him its Wollaston medal. At first it may seem strange that this honour, which had been well earned at least a quarter of a century back, should have been so long deferred. But for the long period of two-and-thirty years he had never ceased to be a member of the Council of the Society, and, as such, could not receive the medal. It was now the accident of his retirement from the Council as one of its senior members that enabled the Society to present him with its highest distinction. Three

circumstances lent an interest to the ceremony. He could
not but recall the time, more than forty years back, when
Dr. Wollaston, whose medal he now received, welcomed him
with friendly words and wise advice into the scientific circles
of London.   Again, by a curious coincidence, the President
of the Society from whose hands he received the medal was
none other than his old friend and colleague, Professor A.
C. Ramsay, who, in bestowing the honour, said,—"Perhaps on this occasion I may be pardoned for recalling the
memory of a time I well remember, when, of all the geologists of weight, you, sir, were the first who held out the
hand of fellowship to me as a young man, when four-and
twenty years ago I was struggling to enter into the ranks
of geologists."   Moreover, the day fixed for the ceremony
chanced to be Murchison's seventy-second birthday.   We
can hardly imagine a conjunction of coincidences more fitted
to make a man ponder deeply on the irrevocable certainties
of his past, and the unknown possibilities of his future.

From his own Sovereign, too, he received in his later
years some marks of distinction which he greatly prized.
Having been made in 1863 a Knight Commander of the
Bath, he was in the year 1866 raised to the dignity of a
Baronet.   Nor were similar tokens of recognition wanting
from abroad.   Other European sovereigns conferred titles
upon him.   The Emperor of Russia, son of his old patron
Nicholas I., sent him a friendly and flattering message, with
a gold snuff-box set with diamonds ; and having already received the Prix Cuvier, Murchison, on the death of Faraday,
was chosen to fill the vacancy in the list of the eight foreign
members of the Academy of Sciences in the Institute of
France.

CHAPTER XXIX.

THE CLOSE.

For many years Lady Murchison's health continued to be delicate. Now and then, as above narrated, she seemed almost at the point of death, while, even when comparatively well, her condition rendered great care necessary. It was under such trying circumstances that she had striven for many years to do her part in the social gatherings which made her house in Belgrave Square one of the chief scientific centres of London. At last she died on the 9th February 1869. Bound to her by a tender attachment, and by the respect which her many excellencies of character never failed to inspire, Murchison acknowledged to the last his gratitude to her as the real source of all his scientific work and fame. A long union of mutual help and sympathy was now severed, and he stood alone in the world, wifeless and childless. At first the blow seemed to have fallen so heavily as to paralyse him, but he quickly rallied. One of his first efforts was to pen the following sketch of his wife, which, though its details have already been, for the most part, given in these chapters at greater length, shows at least the writer's deep affection and gratitude to his life-long companion :—

"*March* 3, 1869.—MY DEAR GEIKIE,—. . . I will give you
a few words upon my beloved wife's influence over me for
whatever good I may have done in the walks of science.   In
the year 1815, the battle of Waterloo having submerged all
my ambition, as well as that of the great Napoleon, and
seeing no 'avenir,' I fell in love with Charlotte Hugonin,
and married her when I was but twenty-three, she being
three and a half years older than myself.   Her father, the
old General of the 4th Dragoons, was a remarkably intelligent man, and a fair astronomer ; her mother a most skilful
florist and botanist, so that their only daughter was brought
up under the most auspicious circumstances.   She was a
good sketcher of scenery, having been taught by the famous
Paul Sandby. . . . Passing our first winter in Hants, I
naturally profited much by the instructions of herself and
parents in all natural-history subjects, and we then prepared
ourselves for a foreign tour in France, the Alps, and Italy,
which we undertook in the spring of 1816.   I was then a
prodigious walker, and more than once did on foot distances
in one day which occupied Swiss horses and carriages two
days, my wife (when practicable) accompanying me on
horseback, and always making me recognise the numerous
wild-flowers peculiar to certain rocks, altitudes, and mountains.   Passing the winter at Genoa (1816-1817), we became
together good Italian scholars, and acquainted with all the
fine art of that noble city.   In the spring we travelled to
Rome, and were so enchanted that we stayed on till the 29th
June, a most imprudent act, by which my wife, who had been
riding late in the Borghese Gardens, caught a malarian
fever, by which I very nearly lost her, and the malady hung
about her through her long life.   We passed the summer at

Naples, where she made numerous coloured sketches, and where we enjoyed the charming society of that capital in those days, and all the enchanting scenery of the environs.

" In our journey homewards we visited every remarkable object, and of all the sights, pictures, and statues I have note-books full, for all that time I was a virtuoso and dilettante.

" Reaching England in 1818, I took to a country residence in the house of my wife's grandfather, recently deceased— the old General who led the 4th Dragoons in one day from Canterbury to the Borough of London, and dispersed the mob in the evening,—a service which George the Third always spoke of with admiration and gratitude. I then gave myself up recklessly but jovially to a fox-hunting life. It was during the years 1818-22 (three in the north country, and two seasons at Melton Mowbray) that my wife was always striving to interest me in something more intellectual than the chase, and began to teach herself mineralogy and conchology. Just at that time (1823), I happened to meet Sir Humphry Davy at Mr. Morritt's of Rokeby, the eminent scholar, and friend of Walter Scott; and Sir Humphry, seeing how my wife was striving to lead me into other paths, gave me words of encouragement, which, coming from so good a man, flattered me, and led me to try and acquire some knowledge of science. He saw that as a sportsman I had a quick and clear eye for a country, and that with most mountain forms and features I was already acquainted, and so he stimulated me to sell my horses, settle for the winter in London, and attend the chemical and mining lectures of the Royal Institution.

" Thus my fate was decided. My first burst of enthusiasm when I got my lessons from Buckland, Greenough, Fitton,

Warburton, Webster, and others, was unbounded, and I was then, to the great delight of my wife, another man.

" Immediately I had acquired this taste my wife and self resolved to explore for ourselves (1825) the whole southern coast of England, from the Isle of Wight inclusive, where all our home phenomena were repeated, to Land's End. For this purpose I had a nice little pair of horses, a light carriage, and with saddles strapped behind to use the nags for riding when at any centre of attraction. At some places we examined the cliffs in boats, she never failing to make good sketches. When we reached Lyme-Regis, she being rather fatigued, I left her to recruit there and amuse herself, and become a good practical fossilist, by working with the celebrated Mary Anning of that place, and trudging with her (pattens on their feet) along the shore ; and thus my first collection was much enriched.

" The year 1826 was very dear to me, for then it was that my good wife accompanied me to the Yorkshire coast, and made many a sketch of cliff and fossil for me, and thence travelled with me to Brora, and various parts of the Hebrides. There also we had our little horses ; and many were the rides she took in Skye and many other places, where wheels could not go. Then it was, too, that she found the *Ammonites Murchisoniæ*, and many other fossils, first described by Sowerby.

" When we were boating it along the shores of Arran in 1826, old Ronald Macalister, the guide of Jameson, who, with a little boy to help, was our only boatman, got quite ' fou ' with his too frequent drams, and as I was thus obliged to row back most of the way from Loch Ranza to Brodick, he amused my poor wife much when, sitting coiled up in the bow, he kept saying, ' Noo, I'm the Laird.'

" In 1828 she saved my life by her energy in treating me for a violent fever caught at Frejus, in the south of France, when walking with Lyell.

" To go on narrating, even in this superficial way, all her adventures with me—all the happy hours we have spent together at the tables of Cuvier, Brongniart, and many eminent men of France, Italy, and Germany, is impracticable in a hurried letter, in scribbling off which all the deepest wells of my heart are opened out.

" In 1830 she explored a large portion of Germany, and the Austrian and Tyrolese Alps, with me, and was with me at Vienna when George the Fourth died, and when we dined with Lord Cowley, at whose table Prince Metternich was quite pleased with her conversation. The Archduke John of Austria (a highly intellectual man) was also her great admirer.

" Need I say that in originating and completing the Silurian classification, from 1830 to 1838, she was very frequently at my side.

" Then as years rolled on, and she became more infirm, she necessarily could not encounter long journeys, and to bivouac with me in Russia and Siberia was out of the question ; but the intense interest she took in my exploration of these regions is best told in her affectionate and dear letters to me, which I cherish, and which buoyed me up through all my difficulties and illnesses in the hope of rejoining her.

" Still feeling that though she was unequal to a Russian campaign she was equal to one in the Alps and Italy, she again set out with me in 1847, when, making the round of Germany, we were together at the last meeting of the Scienziati Italiani at Venice, where all the leading men of science paid her the most marked attention, including Leopold von Buch, Robert Brown, Carl Ritter, etc.

"After this my younger friends came upon the scenes of life. You as well as any one know how much my good, generous, and kind wife did, when other physical powers failed, to cheer and encourage all those who were striving to advance natural knowledge.—Yours ever sincerely,

"Rod. I. Murchison.

"*P.S.*—I may tell you that when the Prince Consort, many years ago, called in Belgrave Square to see my great Russian vase, which my wife was showing to him before I could get to the spot (for I was in my dressing-gown), H. R. Highness said to me, 'I know who made you a geologist.' It would appear that our gracious Queen has always recollected this fact, for not only on all occasions has Her Majesty been most attentive and kind, but on this last melancholy occasion she has specially condoled with me in most touching terms. Among great people the Queen of Holland and Comte de Paris have also sympathized with me, and I have at least 150 letters of condolence, some of them beautifully expressed. Those of yourself and your good Surveyors shall all be bound up with those of other friends in the album of souvenirs of Lady M."

Of all the letters of sympathy which Murchison received at this time, that which probably touched him most was one from Sedgwick. For years past there had been no intercourse between them. Murchison had given up the attempt to conciliate his friend, but never ceased to deplore their estrangement. Only about a month before Lady Murchison's death letters had passed between them relative to a copy of a new edition of 'Siluria' which had been presented by its author to Sedgwick. Murchison had eagerly embraced that occasion once more to try to heal the breach but in vain. The hand of death, however, now touched a chord in Sedgwick's heart which for many a year his old friend had sought in vain to reach. The sight of that former friend and comrade bowed by the deepest sorrow of a man's life, and the recollection of the kindly ways of her who was gone, broke down all barriers, leaving his manly piety and generous feelings to gush forth in the following letter—fitting conclusion to the intercourse which had brightened so many years of their lives, and had linked the names of Sedgwick and Murchison so honourably together in the history of science :—

"Sunday Morning, Feb. 21*st,* 1869.

"DEAR SIR RODERICK MURCHISON,—I did not wish to intrude myself on your sorrows too soon. Indeed, such has been my life of solitude for the last two months, that incidents of the greatest interest to my heart have more than once passed away for a full week or ten days before their report reached me. You will, I know, believe me when I say that the first news of your beloved wife's death filled me with very deep sorrow. For many many years Lady Murchison was one of the dearest of those friends whose society formed the best charms of my life. How often was I her guest! How often have I experienced her kind welcome, and been cheered and strengthened by it! In joy or in sorrow she was my kind and honoured lady friend.

"And have I forgot those bright and, to me, thrice happy days when she and you were my guests at Cambridge ? The present has comparatively little for me now. Hope I have for the future, and I trust that God will give it to me

in the last hours of this world's life whenever they may
come.   But an old man necessarily has his thoughts carried
to the past. But, oh ! how many of the dearest and sweetest
remembrances of my life are now blended with clouds of
sorrow ! It must be so. It is nature's own law. May
God teach you to bear your sorrow like a man. Of this I
have no fear ; but more than this, may His grace be given
you to bear it like a Christian. This sustaining power is
His precious gift, and it must be humbly sought for, by prostration of heart, while under God's afflicting hand. May He
give you the comfort of Christian hope ; compared with it
all other comfort vanishes into mid-air. And if it indeed
be given you, sorrow will lose its bitterness, and even be
tempered with joy.

" I was much affected, and grateful too, when on Friday
last I received what I thought a letter from Frank Buckland
containing his biographical notice of Lady Murchison.   I
immediately wrote to thank him for it. But last night, on
looking at the envelope, I found from the crest and your
name on one corner of it that the 'Souvenirs' had come
from yourself, and now I send you the thanks of an honest
heart for this great kindness. May God bless you and
restore you to comfort !

" If I live four weeks more I shall finish my 84th year.
For more than six months I have been lame from some
chronic affection of the tendon and integuments of my left
knee. I can bear no walking exercise, though I hobble on a
stick as far as my museum. For two months I have greatly
suffered from a combined attack of gout and bronchitis. The
gouty diathesis remains, but the bronchitis has given way,
leaving, however, behind it a painful affection of my eyes,

for which I am to consult a London oculist so soon as I can safely undertake the journey. I generally dictate my long letters to my servant, but in writing this letter of sympathy, addressed to you in your hours of sorrow, I could not find in my heart to use the pen of an amanuensis.

" My eyes are now very angry.—I remain, in all Christian sympathy and good-will, faithfully yours,

" A. SEDGWICK."

She could have been no ordinary woman whose memory drew such an encomium from such a man. Her influence upon the career of her husband was not her only title to the grateful recollection of lovers of science. To the courteous bearing of a cultivated woman she added a brightness of conversation, an intelligence, and a range of knowledge which gave her a peculiar charm, and enabled her to please people of the most varied tastes and acquirements.[1] To her presence the success of her husband's social gatherings was largely due, and there can be little doubt that these gatherings, by commingling students of science with statesmen and politicians, men of letters and men of rank, helped to give science and its cultivators a better hold on the sympathy and good-will of the rest of society.

Though Murchison came to Jermyn Street as of old, and wrote letters or transacted other business there, and though he soon renewed his energy at the Geographical Society, he did not seem ever really to recover from the shock of his wife's death. In these public matters, as well as in the socialities of private life, he appeared, indeed, to

---

[1] Alexander von Humboldt used to describe her as "la spirituelle Lady Murchison."

mingle as much as ever. But though Lady Murchison had been more or less an invalid for some years, she could share up to the last in her husband's cares and interests. Her death broke up therefore the daily intercourse and habits of more than half a century. From such a blow it was hardly to be expected that he should wholly rally.

Of the incidents in Sir Roderick's life after this event no special mention need be made. They continued to be much what they had been for some years before. Among them however there is one which remains to be noticed, as the last effort made in his lifetime for the advancement of the science to which he had given his unremitting energies, and from which in return he had reaped so large a measure of renown—the founding of a Professorship of Geology in the University of Edinburgh. When, on the death of Edward Forbes, the Chair of Natural History became vacant in that University, the Crown, in whose hands the patronage lay, having regard to the enormous strides made by the various sciences which had been taught under that title, anticipated the probable necessity of dividing the Chair into at least two. Various movements were subsequently made to induce the Government to carry out this idea and make the subject of Geology and Mineralogy the business of a new and distinct Professorship. Murchison had taken a part in the negotiations which, however, proved unsuccessful.

In the summer of 1869 he went north for the last time to Scotland to get a little rest, and once more to breathe the air of his native Highlands. The time of his arrival in Edinburgh happened to coincide with the graduation day at the close of the summer session of the University. With

some difficulty he was persuaded to remain for the ceremony, and to receive himself on that occasion the degree of
LL.D. It so chanced that the announcement of a memorial
Fellowship for the encouragement of Geology and Palæontology, recently instituted in the University by the friends
of the late Dr. Hugh Falconer, was made at the same time.
In a short speech on the occasion he by a slip of the tongue
alluded to the *Chair* of Geology which had just been made,
but instantly correcting himself, he added that he hoped
there would before long be a Chair too. The way in which
this was said seemed to indicate that he already meditated
founding the Chair himself.

In the summer of the following year, Professor Allman,
who had succeeded Forbes, resigned his appointment, making
the Natural History Chair once more vacant. Murchison
then determined to carry into effect in his lifetime a proposal
which he had already provided for in his will. He applied
to the Government to divide that Chair into one of Natural
History or Zoology, and one of Geology and Mineralogy, and
offered to provide more than half (£6000) of the endowment
of the new Professorship. Eventually this proposal was
accepted, and on 10th March 1871 a Royal warrant was
issued founding the Chair and appointing the Professor.[1]

It was the first Professorship which had been founded
in any Scottish University for the special teaching of Geology, and there was a peculiar fitness in the fact that it
should have been founded by one who had himself done so
much for Scottish geology. In his remembrance, he wished

[1] Sir Roderick had at different times during the previous ten years
spoken to myself about the desirability of getting this Professorship
established in Edinburgh. At his request the presentation was now given
to me.

it to be known in all time coming as the "Murchison Professorship."

Before the negotiations connected with this matter were brought to a close, the veteran geologist, still vigorous alike in mind and in body, was struck down by paralysis. On the morning of 21st November 1870, while dressing, he had a shock which deprived him of the use of his left side. For a time his life was in some danger, but in a few weeks he so far rallied as to be able to be wheeled into his library. As his speech had only been slightly affected, and as he seemed for a while to be regaining the use of his disabled left hand, he spoke in a cheerful way of his probable recovery. Hence he continued to take a lively interest in current affairs, dictated his correspondence, saw his old friends when they came to inquire for him, and was taken out almost daily in his carriage. In the spring of 1871 he prepared his last Anniversary Address to the Geographical Society, dictating it to his nephew. He knew it would be his last, for even should he recover from the attack, he felt that he could never again take the same active part in life. He had been altogether fifteen years President of the Society, had given sixteen anniversary discourses, and had seen its membership increased from 600, when he was first called to preside, to the large number of 2400. He now resigned the Chair, and with an expression of pride in the success which had attended his efforts to promote the Society's interests, handed the seals of office to his successor. As a recognition of his services, both to the Society itself and to the cause of Geography all over the world, his associates gave him the Founder's medal.

This and other tokens of esteem and grateful recollection cheered the invalid. Unable to walk or stand, he could no longer resume his place at the School of Mines, or at any of the many meetings where he used to be so constant an attendant. But driving about in London, receiving visits from his more intimate friends, and reading, as he did, a good deal, he by no means felt himself cut off from all interest and participation in what was going on in the world. Throughout the summer he continued in this condition, making no visible progress towards convalescence, but yet retaining so much vivacity, and looking so well, that it seemed as if he might yet live for some time to come. He carried on his correspondence, usually by the help of an amanuensis, but sometimes with his own hand, down to the month of August. Some of his letters to myself, written even as late as the early part of that month, though not suitable for quotation here, show little change in the keenness of his interest in the progress of geology, of the British Association, of the School of Mines, the Geological Survey, and other matters with which he had long been so closely connected.

The malady, however, made great progress in the autumn. He had repeatedly expressed a wish to see me, and at the end of September I rejoined him. The lapse of a few weeks had produced a marked and sad change. His speech had become so affected that even his nephew, who assiduously watched him daily, could not make out what he said. His face brightened with the old friendly smile as I sat down beside him for the last time. There was something which he wished to say, but he tried in vain to express it in words. He then had recourse to the pencil, which for a week or two had served to make his wants known to those about him. But the fingers could no longer form any intelligible writing. His eyes filled with tears, and he sank back into his chair.

The end, now plainly near, though sad, could not but be welcome. He had never all his life been given to speaking on religious subjects, but he seemed to enjoy the Psalms and other passages of Scripture as read to him. His nephew asked him if he felt perfectly happy, and received in return a smile and an affirmative pressure of the hand. In the middle of October, in the course of his usual drive, he caught cold. An attack of bronchitis followed, under which, on the 22d of the month, after a lapse of only three days, he quietly and almost imperceptibly passed away.

On the 27th October the remains of the old soldier and geologist were laid beside those of his wife in the Brompton Cemetery. A goodly company of mourners followed them to the grave, including representatives from the varied circles of life and activity where he had moved for so many years, and from which his presence would long be missed. The Queen and the Prince of Wales testified their respect by sending their carriages to join the funeral procession. Among those who walked bareheaded behind the bier the most conspicuous form was that of the Prime Minister, Mr. Gladstone. And thus, amid the deep regrets of his personal friends, and with the respect and esteem of every rank and condition of men, the earth closed upon all that was mortal of Roderick Impey Murchison.

Here the task of the biographer might fitly end. Yet

not without reluctance can he lay down the pen. For many months it has seemed to him as if he had been living again with the friend whose life and work he has been tracing, and from whom the completion of these pages brings as it were a final and irrevocable parting. Here then at the close he would ask what that friend was, and what he did, that his death should have called forth so general an expression of regret. Looking back upon the foregoing narrative, we can perceive that the services by which Murchison earned the esteem and grateful recollection of his fellow-men were twofold. There was first the value of his scientific work, and secondly, the influence of his personal character.

1. It is probably still too soon to attempt an estimate of the actual and lasting contributions made by Murchison to science. But as to the general nature and tendency of his work there can be little diversity of opinion. He was not gifted with the philosophic spirit which evolves broad laws and principles in science. He had hardly any imaginative power. He wanted therefore the genius for dealing with questions of theory, even when they had reference to branches of science the detailed facts of which were familiar to him. The kind of opposition he offered to the views of the evolutionists, and to the doctrines of those who gainsaid his own favourite faith in former convulsions of nature, showing as it did a warmth of antagonism rather than an aptitude for coherent and logical argument, may be cited as evidence of this natural incapacity as well as of the want of early training in habits of accurate scientific reasoning.

But though his name may never be inscribed among those of the recognised magnates in science who are both consummate observers and philosophic reasoners, and who mould the character of science for their own and future times, he will ever hold a high place among the pioneers by whose patient and sagacious power of gathering and marshalling facts new kingdoms of knowledge are added to the intellectual domain of man. He was not a profound thinker, but his contemporaries could hardly find a clearer, more keen-eyed, and careful observer. He had the shrewd-ness, too, to know wherein his strength lay. Hence he seldom ventured beyond the domain of fact where his first successes were won, and in which throughout his long life he worked so hard and so well.

In that domain he had few equals. The patient in-dustry and untiring perseverance with which year after year he returned to the attack of the old grauwacke rocks of his Silurian region form a lesson of hope and encouragement to all students who seek to advance our knowledge of the earth. His Silurian System, in its original elaboration, and in its subsequent extension to the rocks of other countries, may be taken as the type of his scientific work, as it cer-tainly constitutes the ground on which his name will most securely rest. Theories and speculations which now seem firmly established may eventually be swept away before the onward march of research, but such solid contributions of fact as the details of the Silurian System will remain as part of the common stock of knowledge out of which new theories and speculations will be evolved. That system embodied the results of such patient toil as enables a tra-veller to bring a new and unexplored country to the know-ledge of the rest of the world. Murchison's labours among the older rocks stood indeed to geology in a relation not unlike that which his friend Livingstone's work in Africa bore to geography. Round these rocks there had gathered some share of the mystery and fable which hung over the heart of Africa. And he dispelled it not by intuitive genius, but by plodding and conscientious toil, directed by no common sagacity, and sustained by an indomitable courage. For this service his name will be held in lasting and honoured remembrance.

It was in the province of palæozoic geology that Murchison exerted his chief influence upon the progress of science. But, as we have seen, there were other spheres of work wherein he did good service, though its value may be even less easily appraised. No man could be so long and so actively concerned in the direction of some of the leading scientific Societies of his day without materially affecting the advancement of the studies to which these Societies were devoted. To none of its founders and promoters, for example, did the British Association owe more than to him. His work, too, at the Geographical Society, was directly related to some of the best achievements of modern geography. But in this and similar cases his scientific endowments were probably less concerned than his personal character, to which we may now, and lastly, turn.

2. A man's face and figure afford usually a good indication of the general calibre of the spirit which lodges beneath them. The picture which rises to the mind when one thinks of Murchison is that of a tall, wiry, muscular frame, which still kept its erectness even under the burden of almost fourscore years. It seemed the type of body for an active geologist who had to win his reputation by dint of hard climbing and walking almost as much as by mental power. It was moreover united in his case with

a certain pomp or dignity of manner which at one time recalled the military training of the Peninsular days, at another the formal courtesy of the well-bred gentleman of a bygone generation. No learned body or business meeting or anniversary dinner could well be presided over by one who possessed in a greater degree the preliminary and often very useful advantage of a commanding presence. The dignity, however, was blended with a courtesy and kindliness of manner which usually conciliated even those who might have been most disposed to object to any assumption, or appearance of assumption, of authority on his side. So he moved among his fellows as a leader under whom, in the conduct of affairs, his comrades, even when confessedly his own superiors in mental power and scientific achievement, gladly, and indeed instinctively, ranged themselves.

Fortunately his social position and wealth were such as to give him the full use of these personal advantages. Like Sir Joseph Banks, he made his house in Belgrave Square one of the centres where the most truly representative gatherings of men could be met with. Ministers of State, men of rank, of science, of literature, of art, and of travel mingled there together, and came to see and know each other. And yet in the midst of this general intercourse Murchison never lost sight of his scientific position. His guests, too, though they saw him to be a man of the world, were in various indirect ways reminded that he took a pride in his science. It may be hardly possible to estimate the value of this influence. But assuredly among the causes which have helped, during the last thirty years, to give science and its votaries a firmer hold upon society, and especially the higher classes of society, we must count as by no means unimportant the liberal way in which Murchison displayed his excellent social qualities.

But above and beyond this mere external aptitude for a place of eminence, Murchison had many higher claims to such a position. Foremost we should place his vigorous energy, his unwearied and almost restless activity. He seemed never to be without a definite and well-planned task. When his hands were fullest of his own work, he appeared to have almost unlimited time for assisting the labours of others, and co-operating with them for the general advancement of knowledge. That he could do this in so great a degree arose not only from his capacity for work, but from a certain method and orderliness of mind which characterized him in every period and phase of his life. The spirit which led him in early days to tabulate the deeds and fate of his respective hunters was the very same which guided him through the labyrinths of grauwacke, and prompted his exertions for the welfare of the Geographical Society. Men could not but respect one who, while doing so much honest independent work himself, was ever ready to take his place with others in efforts for the general good.

Another leading feature of his character, and an element which largely aided his success, was shrewd common sense and knowledge of the world. We see this feature conspicuously manifested in all his scientific undertakings, where he derived more help from it than some of his contemporaries did from undoubted genius. He never allowed himself to be led astray from the track which he was patiently and ploddingly following by any will-o'-the-wisp in the shape of speculation or theory; nor in the management of affairs did he bind himself to lofty and impracticable principles. He took men and things as he found them, and tried to work upon them by firmness or concession, as seemed most likely to aid his object. Now and then, when provoked by opposition, he manifested a certain impatience, and even imperiousness of manner, which provoked rather than conciliated. Nevertheless, in the tact which enables a man to manage his fellows successfully for many years, he had few rivals. He showed it in the conduct of the various learned Societies of whose governing bodies he was a member. But nowhere did he display it more conspicuously than in the way in which he gained from different Ministries a recognition of the claims of scientific discovery. Probably no man had so much influence with the Governments of his day, and no man more honourably, persistently, and courageously used it.

There was still another characteristic which secured to Murchison the esteem as well as the respect of his fellow-men—his thorough kindliness and goodness of heart. Separate instances of this have been given in the foregoing narrative, but it was a feature which showed itself all through his life. Many a humble fellow-worker in science did he encourage and materially assist. When he had given the right hand of friendship to a man he stuck to him, even in the face of baseness and ingratitude. The devotion; indeed, with which he espoused the cause of a friend, had, at times, something altogether chivalrous about it. Of this, the instance which will naturally rise to most men's minds is his hearty and energetic devotion to Livingstone, whose interests he so entirely identified with his own.

With the recollection of these features which go to make up the picture of what Murchison was, there must needs mingle some slight remembrance of his foibles of character. But if this narrative of his life has been as faithful as its writer has wished that it should be, these superficial weaknesses have already appeared and need not be touched on here. Rather let us carry with us through the rest of life the lessons which the other and dominant features of his character and work may teach—his persevering industry, his readiness to be helpful, his loyalty to a friend, and, above all, his life-long and entire devotion to the advancement of knowledge. It will, perhaps, be many a day before another man arises to fill among us the honourable and useful place from which we shall long miss the presence of Roderick Impey Murchison.

LIST OF SIR RODERICK MURCHISON'S PUBLISHED WRITINGS.

1825. Geological Sketch of the North-Western Extremity of Sussex, and the adjoining parts of Hants and Surrey.
Geol. Soc. Trans., second series, ii. 97.

1827. On the Coal-field of Brora, in Sutherlandshire, and some other Stratified Deposits in the North of Scotland.
Geol. Soc. Trans., second series, ii. 293.
Supplementary Remarks on the Strata of the Oolitic series and the rocks associated with them in Sutherland, Ross, and the Hebrides.
Geol. Soc. Trans., second series, ii. 353.

1828. On the Geological relations of the Secondary Strata in the Island of Arran; by A. Sedgwick and R. I. M.
Geol. Soc. Proc., i. 41. *Geol. Soc. Trans., second series;* iii. 21.
On the Old *Red* Conglomerates and other Secondary Deposits on the North Coast of Scotland, by A. Sedgwick and R. I. M.
Geol. Soc. Proc., i. 77. *Geol. Soc. Trans., second series,* iii. 125.
On the Excavation of Valleys, as illustrated by the Volcanic Rocks of Central France, by C. Lyell and R. I. M.
Geol. Soc. Proc., i. 89.

1829. On the Tertiary and Secondary Rocks forming the Southern Flank of the Tyrolese Alps, near Bassano.
Geol. Soc. Proc., i. 137.

1829. On the Bituminous Schists and Fossil Fish of Seefeld, in the Tyrol.
> *Geol. Soc. Proc.,* i. 139.

On the Tertiary Deposits of the Cantal and their relation to the Primary and Volcanic Rocks, by C. Lyell and R. I. M.
> *Geol. Soc. Proc.,* i. 140. *Ann. Sci. Nat.,* xviii., 1829, p. 173.

On the Tertiary Fresh-water Formations of Aix, in Provence, including the Coal-field of Fuveau, by C. Lyell and R. I. M.
> *Geol. Soc. Proc.,* i. 150.

On the Tertiary Deposits of the Vale of Gosau, in the Salzburg Alps, by A. Sedgwick and R. I. M.
> *Geol. Soc. Proc.,* i. 153. [*See Geol. Soc. Trans., second series,* iii. 301.]

On the Tertiary Formations which range along the flanks of the Salzburg and Bavarian Alps, by A. Sedgwick and R. I. M.
> *Geol. Soc. Proc.,* i. 155.

1830. On the Fossil Fox of Œningen, with an account of the Lacustrine Deposit in which it was found.
> *Geol. Soc. Proc.,* i. 167. *Geol. Soc. Trans., second series,* iii. 277.

On the Tertiary Deposits of Lower Styria, by A. Sedgwick and R. I. M.
> *Geol. Soc. Proc.,* i. 213.

A Sketch of the Structure of the Austrian Alps, by A. Sedgwick and R. I. M.
> *Geol. Soc. Proc.,* i. 227. *Geol. Soc. Trans., second series,* iii. 301.

1831. Supplementary Observations on the Structure of the Austrian and Bavarian Alps.
> *Geol. Soc. Proc.,* i. 249. *Geol. Soc. Trans., second series,* iii. 301.

Notes on the Secondary Formations of Germany as compared with those of England.
> *Geol. Soc. Proc.,* i. 325.

1832. Presidential Address to Geological Society.
> *Geol. Soc. Proc.,* i. 362.

On the Structure of the Cotteswold Hills and country around Cheltenham, and on the occurrence of stems of

fossil plants in vertical positions in the sandstone of the Inferior Oolite of the Cleveland Hills.
Geol. Soc. Proc., i. 388.

1833. Presidential Address to Geological Society.
Geol. Soc. Proc., i. 438.
On the Sedimentary Deposits which occupy the western parts of Shropshire and Herefordshire, and are prolonged from N.E. to S.W. through Radnor, Brecknock, and Caermarthen shires, with descriptions of the accompanying rocks of intrusive or igneous characters.
Geol. Soc. Proc., i. 470.

1834. On the Old Red Sandstone in the counties of Hereford, Brecknock, and Caermarthen, with collateral observations on the Dislocations which affect the North-west Margin of the South Welsh Coal-basin.
Geol. Soc. Proc., ii. 11.
On the Structure and Classification of the Transition Rocks of Shropshire, Herefordshire, and part of Wales, and on the Lines of Disturbance which affected that series of deposits, including the Valley of Elevation of Woolhope.
Geol. Soc. Proc., ii. 13.
On Fresh-water Limestone between the seams of Coal in the neighbourhood of Shrewsbury.
Phil. Mag., third series, iv. 158.
On the Gravel and Alluvial Deposits of those parts of the counties of Hereford, Salop, and Worcester, which consist of Old Red Sandstone; with an account of the Puffstone or Travertin of Spothouse, and of the Southstone Roch near Tenbury.
Geol. Soc. Proc., ii. 77.
On certain Trap-rocks in the counties of Salop, Montgomery, Radnor, Brecon, Caermarthen, Hereford, and Worcester, and the effects produced by them upon the Stratified Rocks.
Geol. Soc. Proc., ii. 85.
On the Upper Greywacke Series of England and Wales. Table of the Order of Stratified Deposits which connect the Carboniferous Series with the older slaty rocks in the counties of Salop, Hereford, etc.
Edin. New Phil. Journ., xvii. 365.

1835. On an Outlying Basin of Lias on the borders of Salop and

Cheshire, with a short account of the Lower Lias between Gloucester and Worcester.
Geol. Soc. Proc., ii. 114.

1835. A general view of the New Red Sandstone Series in the counties of Salop, Stafford, Worcester, and Gloucester.
Geol. Soc. Proc., ii. 115.

On certain Coal tracts in Salop, Worcestershire, and North Gloucestershire.
Geol. Soc. Proc., ii. 119.

On certain lines of Elevation and Dislocation of the New Red Sandstone of North Salop and Staffordshire, with an account of Trap-dykes in that formation at Acton Reynolds, near Shrewsbury.
Geol. Soc. Proc., ii. 193.

On the Silurian System of Rocks.
Phil. Mag., third series, vii. 46.

On the Silurian and Cambrian Systems, exhibiting the order in which the older sedimentary strata succeed each other in England and Wales, by A. Sedgwick and R. I. M.
Brit. Assoc. Rep., 1835, pt. 2. 59.

On the Recent Discovery of Fossil Fishes (*Palæoniscus catopterus*, Agass.) in the New Red Sandstone of Tyrone, Ireland.
Geol. Soc. Proc., ii. 206.

1836. On the Geological Structure of Pembrokeshire, more particularly on the extension of the Silurian System of rocks into the coast cliffs of that county.
Geol. Soc. Proc., ii. 226.

On the Gravel and Alluvia of South Wales and Siluria as distinguished from a northern drift covering Lancashire, Cheshire, North Salop, and parts of Worcester and Gloucester.
Geol. Soc. Proc., ii. 230.

On the Silurian and other rocks of the Dudley and Wolverhampton Coal-field, followed by a Sketch proving the Lickey Quartz-rock to be of the same age as the Caradoc Sandstone.
Geol. Soc. Proc., ii. 407.

On the supposed existence of the Lias Formation in Africa.
Geol. Soc. Proc., ii. 415.

On the Ancient and Modern Hydrography of the river Severn.
Brit. Assoc. Rep. 1836, *Sect.*, p. 88.

1836. A Classification of the old Slate-rocks of the North of
Devonshire, and on the true position of the Culm
deposits in the central portion of that county, by A.
Sedgwick and R. I. M.
Brit. Assoc. Rep., 1836, *Sect.*, p. 95.
Description of a raised Beach in Barnstaple Bay, on the
north-west coast of Devonshire, by A. Sedgwick and
R. I. M.
Geol. Trans., second series, v. 279. *Geol. Soc. Proc.*,
ii. 441.

1837. On the Physical Structure of Devonshire, and on the Subdivisions and Geological Relations of its old stratified
deposits, by A. Sedgwick and R. I. M.
Geol. Trans., second series, v. 663. *Geol. Soc. Proc.*, ii.
556.
On the Upper Formations of the New Red System in
Gloucestershire, Worcestershire, and Warwickshire,
showing that the red (saliferous) marls, with an included band of sandstone, represent the Keuper or
" marnes irisées," and that the underlying sandstone of
Ombersley, Broomsgrove, and Warwick is part of the
" Bunter Sandstein," or " grés bigarré," of foreign
geologists, by R. I. M. and H. E. Strickland.
Geol. Trans., second series, v. 331. *Geol. Soc. Proc.*, ii.
563.
On the Fishes of the Ludlow Rocks.
Brit. Assoc. Rep., 1837, *Sect.*, p. 91.

1838. Notice of a Specimen of the Oar's Rock, nine miles south
of Littlehampton.
Geol. Soc. Proc., ii. 686.
Description of a Self-registering Thermometer and Baro-
meter invented by the late J. Coggan.
Royal Soc. Proc., iv. p. 72.
On the Silurian System of Strata.
Brit. Assoc. Rep., 1838, *Sect.*, p. 80.
Address to the British Association (Newcastle).
Brit. Assoc. Rep., 1838, p. xxxi.

1839. On the Classification of the Older Rocks of Devonshire
and Cornwall, by A. Sedgwick and R. I. M.
Geol. Soc. Proc., iii. 121. *Phil. Mag.*, xiv. 242.
Supplementary Remarks on the " Devonian " System of
Rocks, by A. Sedgwick and R. I. M.
Phil. Mag., xiv. 354.

1839. On the Carboniferous and Devonian Systems of Westphalia.
> *Brit. Assoc. Rep., Sect.*, p. 72.

1840. Sur les roches Devoniennes [type particulier de l'Old
Red Sandstone des géologues Anglais] qui se trouvent
dans le Boulonnais.
> *Bull. Géol. Soc. Paris*, xi. 229.

On the Classification and Distribution of the Older Rocks
of North Germany, etc., by A. Sedgwick and R. I. M.
> *Geol. Soc. Proc.*, iii. 300.   *Geol. Trans.*, vi. 221.

Anniversary Address to the British Association (Glasgow),
by R. I. M. and E. Sabine.
> *Rep. Brit. Assoc.*, 1840, p. xxxv.

On the Fishes of the Old Red Sandstone.
> *Brit. Assoc. Rep.*, 1840, *Sect.*, p. 99.

On the Stratified Deposits which occupy the Northern and
Central Regions of Russia, by R. I. M. and E. de Verneuil.
> *Brit. Assoc. Rep.*, 1840, *Sect.*, p. 105.

1841. Observations géologiques sur la Russie.
> *Soc. Nat. Moscou, Bull.*, 1841, p. 901.

On the Geological Structure of the Northern and Central
Regions of Russia, by R. I. M. and E. de Verneuil.
> *Geol. Proc.*, iii. 398.

Notes on a Section and a List of Fossils from the State of
New York.
> *Geol. Proc.*, iii. 416.

First Sketch of some of the principal results of a second
Geological Survey of Russia, in a letter to Mr. Fischer.
> *Phil. Mag., new series*, xix. 417.   [*Moscow Soc. Nat.
Bull.*, 1841, p. 901.]

Article on Tours in Russian Provinces.
> *Quarterly Review*, No. 134.   (March 1841.)

1842. Presidential Address to Geological Society.
> *Geol. Proc.*, iii. 637.

Results of a second Geological Survey of Russia, by R. I.
M., Keyserling, and Verneuil.
> *Geol. Proc.*, iii. 717.

Inaugural Address at the first general Meeting of the Dudley and Midland Geological Society, Jan. 1842.   London.

On the Salt-steppe south of Orenburg, and on a remarkable Freezing Cavern.
> *Geol. Proc.*, iii. 695.

1842. On the Tchornoi Zem or Black Earth of Central Russia.
 Geol. Proc., iii. 712.
 Memoir on the Geological Structure of the Ural Moun-
 tains, by R. I. M., Keyserling, and Verneuil.
 Geol. Proc., iii. 742.
 On the Distinction between the Striated Surface of Rocks
 and Parallel Undulations dependent on Original Struc-
 ture.
 Brit. Assoc. Rep., 1842, *Sect.,* p. 53.
 On the Glacial Theory.
 Edin. New Phil. Journ., xxxiii., 1842, p. 124.

1843. Observations on the Occurrence of Fresh-water Beds in
 the Oolitic Deposits of Brora, Sutherlandshire; and
 on the British equivalents of the Neocomian system of
 foreign geologists.
 Geol. Proc., iv. 174.
 Presidential Address to the Geological Society.
 Proc. Geol. Soc., iv. p. 65.
 The Permian System as applied to Germany, with col-
 lateral observations on similar deposits in other countries,
 showing that the Rothe-todte-liegende, Kupfer-Schiefer,
 Zechstein, and the lower portion of the Bunter-sand-
 stein, form one natural group, and constitute the upper
 member of the Palæozoic Rocks.
 Brit. Assoc. Rep., 1843, *Sect.,* p. 52.
 On the important Additions recently made to the Fossil
 contents of the Tertiary and Alluvial Basin of the Middle
 Rhine.
 Ibid., p. 55.
 1. The Permian System of Rocks. 2. Theory of the
 Origin of Coals. 3. Lines of Ancient Sea-levels. 4.
 On Mastodontoid and Megatherioid Animals.
 Edin. New Phil. Journ., xxxv. (1843), p. 115.
 On the Geology of Russia and the Ural Mountains.
 Moxon, Geologist, 1843, p. 201.
 A few observations on the Ural Mountains, to accom-
 pany a new map of southern portion of that chain.
 R. Geograph. Soc. Journal, xiii. p. 269.
 Paläozoisches Gebirge : Silurisches, Devonisches, und
 Kohlen System.
 [*L'Institut,* x. 1842.] *Leonard u. Bronn's N. Jahrb.,*
 1843, p. 621.

1844. On the Permian System as developed in Russia and

other parts of Europe, by R. I. M. and E. de Verneuil.

Calcutta Journ. Nat. Hist., vi. 266.   *Geol. Soc. Proc.*,
  iv. 327.   *Geol. Soc. Journ.*, i. 81.   *Bull. Soc. Géol.*,
  France, 1844, i. p. 475.

1844. Address as President of the Geographical Society.
  Journ. Geog. Soc., xiv. p. 45.

Note sur les équivalents du système Permien en Europe,
  suivie d'un coup d'œil général sur l'ensemble de ses
  fossiles, et d'un tableau des espèces, by R. I. M. and E.
  de Verneuil.
  Paris Soc. Géol. Bull., second series, i. 475.

On the Bathymetrical Distribution of Submarine Life on
  the Northern Shores of Scandinavia.
  Brit. Assoc. Rep., 1844, *Sect.*, p. 50.

On the Palæozoic Rocks of Scandinavia and Russia, particularly as to the Lower Silurian Rocks which form
  their true base.
  Brit. Assoc. Rep., 1844, *Sect.*, p. 53.

Ansknelser over Classificationen af de geologiske Lag i
  Overgangs formationen ved Christiania.
  Skand. Naturf. Förhandl., iv. p. 287.

Ueber die allgemeinen Beziehungen zwischen den älteren
  paläozoischen Sedimenten in Scandinavien und in
  Baltischen Provinzen Russlands.
  St. Petersb. Verhandl. Min. Gesell., 1844, p. 190.

1845. Outline of the Geology of the Neighbourhood of Cheltenham, augmented by J. Buckman and H. E. Strickland.
  8vo, London.

Uebersicht der neuesten geographischen und geologischen
  Arbeiten im Russischen Reiche.
  Erman. Archiv. Russ., iv. p. 321.

Ueber fossile Pflanzen.
  [*Athenæum*, 1845.]   *Froriep. Notizen.*, xxxv. col. 343.

On the Palæozoic Deposits of the Basin of Christiania.
  Förhand. Skand. Naturf. Möte. Christiania, 1845.

On the Palæozoic Deposits of Scandinavia and the Baltic
  Provinces of Russia, and their relations to Azoic or more
  ancient crystalline rocks; with an account of some
  great features of Dislocation and Metamorphism along
  their northern frontiers.
  Geol. Soc. Journ., i. 467.   *Geol. Soc. Proc.*, iv. 601.

1846. Geology of Russia and the Ural Mountains, by R. I. M.,

E. de Verneuil, and Count von Keyserling. 2 vols. 4to, London.

1846. On the Superficial Detritus of Sweden, and on the probable causes which have affected the surface of the rocks in the Central and Southern portions of that kingdom.
Geol. Soc. Journ., ii. 349.
A brief Review of the Classification of the Sedimentary Rocks of Cornwall (with opinions on the Gold of Australia).
Trans. Roy. Geol. Soc. Cornwall, vi. 317. *Ann. Mag. N. Hist.,* xix. p. 326. *Edin. New Phil. Journ.,* xliii. 31.
On the Silurian and associated rocks in Dalecarlia, and on the succession from Lower to Upper Siluria in Smoland, Oland, and Gothland, and in Scania.
Q. Journ. Geol. Soc., iii. p. 1.
Presidential Address to the British Association (Southampton).
Rep. Brit. Assoc., 1846.
Additional remarks on the Deposit of Œningen in Switzerland.
Q. Journ. Geol. Soc., iii. p. 54.

1847. On the meaning originally attached to the term " Cambrian System," and on the Evidences since obtained of its being geologically synonymous with the previously established term, " Lower Silurian."
Geol. Soc. Journ., iii. 165. *Edin. New Phil. Journ.,* xliii. 147.
Nouvelles remarques sur la classification des terrains paléozoiques inférieurs.
Paris Comptes Rendus, xxiv. 838.
On the Discovery of Silurian Rocks in Cornwall.
Phil. Mag. xxx. 336. *Ann. Nat. Hist.,* xix. 326.
Introduction to a second Memoir of Captain Vicary on the Geology of parts of Scinde.
Geol. Soc. Journ., iii. 331.
Habitation and Destruction of the Mammoths.
Edin. New Phil. Journ., xl. 344. *Calcutta Journ. Nat. Hist.,* vii. 424.

1848. On the Geological Structure of the Alps, Apennines, and Carpathians, more especially to prove a transition from Secondary to Tertiary Rocks, and the development of Eocene deposits in Southern Europe.

Geol. Soc. Journ., v. 157. *Edin. New Phil. Journ.*, xlvi. 280. *Froriep. Notizen*, x. col. 184. *Phil. Mag.*, xxxiv. 207.

1848. Ueber die silurischen Gesteine Böhmens, nebst einigen Bemerkungen ueber die devonische Gebilde in Mähren.
Leonhard u. Bronn, N. Jahrbuch, 1848, 1.

1849. On the Development of the Permian System in Saxony.
Geol. Soc. Journ., v. 1.
On the Distribution of the Superficial Detritus of the Alps as compared with that of Northern Europe.
Journ. Geol. Soc., vi. 65. *Edin. New Phil. Journ.*, xlviii. 256.
On a Metamorphosis of certain Trilobites, as recently discovered by M. Barrande.
Brit. Assoc. Reports, 1849, *Sect.* 58.
On the Distribution of Gold Ore in the Crust and on the surface of the Earth.
Brit. Assoc. Rep., 1849, *Sect.*, p. 60.

1850. On the Earlier Volcanic Rocks of the Papal States and adjacent parts of Italy.
Geol. Soc. Journ., vi. 281.
On the Distribution of Gold.
Proc. Roy. Inst., 1850.
On the Vents of Hot Vapour in Tuscany, and their relation to ancient lines of Fracture and Eruption.
Geol. Soc. Journ., vi. 367. *Phil. Mag.*, i. 51. *Silliman, Journal*, xi. 199.
Geologia delle Alpi, degli Apennine e dei Carpezi. Treduzione dall' Inglese ed Appendice sulla Toscana dei Professori Paolo Savi e G. Meneghini. Firenze, 8vo.
The Slaty Rocks of the Sichon, or northern end of the Chain of the Forez, in Central France, and on lines of dislocation between the Lower and Upper Carboniferous deposits of France and Germany.
Brit. Assoc. Rep., 1850, *Sect.*, p. 96. [*Geol. Soc. Journ.*, vii. 13.]
Review of the labours of M. Barrande in preparing his important work, "The Silurian System of Bohemia."
Brit. Assoc. Rep., 1850, *Sect.*, p. 97.
Siberia and California.
Quart. Rev., lxxxvii. 397.
On the Origin of the Mineral Springs of Vichy.
Geol. Soc. Journ., vii. 76.

1851. On the Silurian Rocks of the South of Scotland.
    Geol. Soc. Journ., vii. 139.
    On the former changes of the Alps.
    Proc. Roy. Inst., i. 31.   *Edin. New Phil. Journ.*, li. 31.
        Silliman, Journal, xii. 245.
    On the Distribution of the Flint-drift of the South-east of
    England on the flanks of the Weald and over the sur-
    face of the South and North Downs.
    Geol. Soc. Journ., vii. 349.
    On the Scratched and Polished Rocks of Scotland.
    Brit. Assoc. Rep., 1851, p. 66.
    The Slaty Rocks of the Sichon, or northern end of the
    chain of the Forez, in Central France, shown to belong
    to the Carboniferous age.
    Geol. Soc. Journ., vii. 13.

1852. On the Anticipation of the Discovery of Gold in Australia ;
    with a general view of the conditions under which that
    metal is discovered.
    Geol. Soc. Journ., viii. 134.
    A few Remarks on the Silurian Classification.
    Silliman, Amer. Journ., second series, iii. p. 404.
    Communication of Dr. A. Fleming's Memoir on the Salt
    Range of the Punjab.
    Geol. Soc. Journ., ix. 189.
    Address as President of the Royal Geographical Society.
    Journ. Geog. Soc., xxii. p. lxii.
    On the meaning attached by Geologists during the last
    ten years to the term "Silurian System."
    Geol. Soc. Journ., viii. 173.
    A General View of the Palæozoic Rocks.   8vo, London.
    The Silurian System.
    Edin. New Phil. Journ., lii. 355.

1853. On the Basin-like form of Africa.
    Edin. New Phil. Journ., lii. 52.
    On some of the Remains in the Bone-bed of the Upper
    Ludlow Rock.
    Geol. Soc. Journ., ix. 16.
    Address as President of the Royal Geographical Society.
    Journ. Geog. Soc., xxiii. p. lxii.

1854. General Observations on the Palæozoic Rocks of Ger-
    many.
    Brit. Assoc. Rep., 1854, *Sect.*, p. 87.

1854. On a supposed Aërolite or Meteorite found in the trunk of an old willow-tree in the Battersea Fields.
    Roy. Soc. Proc., vii. 421.
Siluria (first edition), 8vo, London.

1855. Additional Observations on the Silurian and Devonian Rocks near Christiania, in Norway.
    Geol. Soc. Journ., xi. 161.
On the Occurrence of numerous Fragments of Fir-wood in the islands of the Arctic Archipelago ; with remarks on the Rock Specimens brought from that region.
    Geol. Soc. Journ., xi. 536.   *Silliman, Journ.*, **xxi.** 377.
On the Relations of the Crystalline Rocks of the North Highlands to the Old Red Sandstone of that region, and on the recent discoveries of Fossils in the former, by Mr. Charles Peach.
    Brit. Assoc. Rep., 1855, *Sect.*, p. 85.
Recherches géologiques dans le nord de l'Écosse.
    Paris, Soc. Géol. Bull., xiii. 21.
On the Discovery of Fossils in the uppermost Silurian Rocks near Lesmahagow, in Scotland, with observations on the relations of the Palæozoic Strata in that part of Lanarkshire.
    Geol. Soc. Journ., xii. 15.

1856. On the Bone-beds of the Upper Ludlow Rock, and base of the Old Red Sandstone.
    Brit. Assoc. Rep., 1856, *Sect.*, p. 70.

1857. Note on the Relative Position of the Strata, near Ludlow, containing the Ichthyolites.
    Geol. Soc. Journ., xiii. 290.
The Silurian Rocks and Fossils of Norway, as described by M. Theodor Kjerulf; those of the Baltic Provinces of Russia, by Prof. Schmidt ; and both compared with their British equivalents.
    Geol. Soc. Journ., xiv. 36.
Address as President of the Royal Geographical Society.
    Journ. Geog. Soc., xxvii. p. xciv.
The Quartz Rocks, Crystalline Limestones, and Micaceous Schists of the North-western Highlands of Scotland, proved to be of Lower Silurian age, through the recent Fossil discoveries of Mr. C. Peach.
    Brit. Assoc. Rep., 1857, *Sect.*, p. 82.

1857. On the Crystalline Rocks of the North Highlands of Scotland.
    Amer. Assoc. Proc., 1857 (pt. 2), p. 57.

1858. Sur une nouvelle classification des terrains de l'Écosse.
    Paris, Soc. Géol. Bull., second series, xv. 367.
    On the Succession of Rocks in the Northern Highlands, from the oldest Gneiss, through Strata of Cambrian and Lower Silurian age, to the Old Red Sandstone inclusive.
    Geol. Soc. Journ., xiv. 501.
    Address as President of the Royal Geographical Society.
    Journ. Geog. Soc., xxviii. p. cxxiii.
    Some results of recent Researches among the older rocks of the Highlands of Scotland.
    Brit. Assoc. Rep., 1858, *Sect.,* p. 94.
    On the Succession of the older Rocks in the northernmost counties of Scotland; with some Observations on the Orkney and Shetland Islands.
    Geol. Soc. Journ., xv. 335.
    On the Sandstones of Morayshire (Elgin, etc.), containing Reptilian Remains; and on their Relations to the Old Red Sandstone of that county.
    Geol. Soc. Journ., xv. 419.

1859. Address as President of the Royal Geographical Society.
    Journ. Geog. Soc., xxix. p. cii.
    On the Commercial and Agricultural Value of certain Phosphatic Rocks of the Anguilla Isles, in the Leeward Islands.
    Agric. Soc. Journ., xx. 31.
    Supplemental Observations on the Order of the Ancient Stratified Rocks of the North of Scotland, and their associated Eruptive Rocks.
    Geol. Soc. Journ., xvi. 215.

1861. Address as President of the Royal Geographical Society.
    Journ. Geog. Soc., xxxi. p. cxi.
    On the Inapplicability of the new term "Dyas" to the "Permian" group of rocks, as proposed by Dr. Geinitz.
    Geologist, v. 4.   *Edin. New Phil. Journ.,* xv. 71.

1862. Thirty years' Retrospect of the Progress in our Knowledge of the Geology of the Older Rocks.
    Silliman, Journ., xxxiii. 1.

1862. Quelques mots sur l'existence de gneiss fondamental ou
laurentian, et sur le développement des dépots de l'age
permien en Bohème.
Paris, Géol. Soc. Bull., xx. 155.

1863. On the Permian Rocks of North-eastern Bohemia.
Geol. Soc. Journ., xix. 297.
On the Gneiss and other Azoic Rocks, and on the super-
jacent Palæozoic Formations of Bavaria and Bohemia.
Geol. Soc. Journ., xix. 354.
Address as President of the Royal Geographical Society.
Journ. Geog. Soc., xxxiii., p. cxiii.
Introduction to Messrs. Gordon and Joass' paper on the
Relations of the Ross-shire Sandstones containing
Reptilian Footprints.
Geol. Soc. Journ., xix. 506.
Address as President of the Geographical and Ethnological
section of the British Association.
Brit. Assoc. Rep., 1863, *Sect.*, p. 126.
Observations upon the Permian group of the North-west
of England.
Brit. Assoc. Rep., 1863, *Sect.*, p. 83.

1864. On the Permian Rocks of the North-west of England
and their Extension into Scotland, by R. I. M. and Pro-
fessor Harkness.
Geol. Soc. Journ., xx. 144.
On the Antiquity of the Physical Geography of Inner
Africa.
Journ. Geog. Soc., xxxiv. 201.
Address as President of the Royal Geographical Society.
Journ. Geog. Soc., xxxiv., p. cix.
Note on the Occurrence of the same Fossil Plants in the
Permian Rocks of Westmoreland and Durham.
Brit. Assoc. Rep. 1864, *Sect.*, p. 59.
Address as President of the Geographical and Ethnological
section of the British Association.
Brit. Assoc. Rep., 1864, *Sect.*, p. 130.
Note on communicating the Notes and Map of Dr. Julius
Haast upon the Glaciers and Rock-basins of New
Zealand.
Geol. Soc. Journ., xxi. 129.

1865. Address as President of the Royal Geographical Society.
Journ. Geog. Soc., xxxv. p. cviii.

1865. Address as President of the Geological section of the
British Association.
Brit. Assoc. Rep., 1865, *Sect.*, p. 41.

1866. Address as President of the Royal Geographical Society.
Journ. Geog. Soc., xxxvi. p. cxviii.
On the parts of England in which Coal may and may not
be looked for beyond the known Coal-fields.
Brit. Assoc. Rep., 1866, *Sect.*, p. 57.
On the reported Discovery of the Remains of Leichhardt
in Australia.
Brit. Assoc. Rep., 1866, *Sect.*, p. 114.

1867. Address as President of the Royal Geographical Society.
Journ. Geog. Soc., xxxvii. p. cxv.
Observations on the Livingstone Search Expedition now
in progress.
Brit. Assoc. Rep., 1867, *Sect.*, p. 126.

1868. Address as President of the Royal Geographical Society.
Journ. Geog. Soc., xxxviii., p. cxxxiii.
Note comparing the Geological Structure of North-western
Siberia with that of Russia in Europe.
Geol. Soc. Journ., xxv. 1.

1869. Address as President of the Royal Geographical Society.
Journ. Geog. Soc., xxxix. p. cxxxv.
Introduction to the Rev. J. M. Joass' Notes on the Suther-
land Gold-field.
Geol. Soc. Journ., xxv. 314.
Observations on the Structure of the North-west High-
lands.
Edin. Geol. Soc. Trans., ii. 18.

1870. Address as President of the Royal Geographical Society.
Journ. Geog. Soc., xl. p. cxxxiii.
Address as President of the Geographical Section of the
British Association.
Brit. Assoc. Rep., 1870, *Sect.*, p. 158.

1871. Address as President of the Royal Geographical Society.
Journ. Geog. Soc., xli. p. cxlvi.

INDEX.

PRINTED BY T. AND A. CONSTABLE, PRINTERS TO HER MAJESTY, AT THE EDINBURGH UNIVERSITY PRESS.

www.ingramcontent.com/pod-product-compliance
Ingram Content Group UK Ltd.
Pitfield, Milton Keynes, MK11 3LW, UK
UKHW040659180125
453697UK00010B/277